GET SMART

MATHS

GET
SMART

THE BIG IDEAS

YOU SHOULD KNOW

MATHS

JULIA COLLINS

Quercus

Contents

Are you a genius?

Check off the topics as you master them in increasing levels of difficulty.

Mind-blowing
- ☐ The Birch and Swinnerton-Dyer conjecture
- ☐ The Riemann hypothesis
- ☐ The Poincaré conjecture
- ☐ The Hodge conjecture
- ☐ The building blocks of symmetry
- ☐ Continuous symmetry
- ☐ The Yang–Mills and mass gap problem
- ☐ The Navier–Stokes equations

Formidable
- ☐ Fermat's Last Theorem
- ☐ Gödel's incompleteness theorem
- ☐ Sizes of infinity
- ☐ The shape of our universe
- ☐ Identifying shapes

Tough
- ☐ Keeping secrets with primes
- ☐ Proofs
- ☐ Turing machines
- ☐ Projective geometry
- ☐ The impossibility of solving equations
- ☐ To the limit
- ☐ Adding up infinity
- ☐ Modelling change
- ☐ Unexpected trigonometry
- ☐ P versus NP and the problem of fast algorithms

Tricky
- ☐ Imaginary numbers
- ☐ Patterns in primes
- ☐ Euclidean geometry
- ☐ Wallpaper patterns and Penrose tilings
- ☐ Non-orientable surfaces
- ☐ Euler's formula and the shape of surfaces
- ☐ Knot theory
- ☐ Topology
- ☐ The mathematics of symmetry
- ☐ Calculus
- ☐ Chaos theory
- ☐ Fractals
- ☐ The four colour theorem
- ☐ The prisoner's dilemma
- ☐ Probability in the courtroom
- ☐ Randomness
- ☐ Paradoxes of probability
- ☐ The prisoner's dilemma

Fundamental
- ☐ Number systems
- ☐ Binary numbers
- ☐ Game theory
- ☐ Irrational numbers
- ☐ Prime numbers
- ☐ Platonic solids
- ☐ The sphere-packing problem
- ☐ Graph theory
- ☐ Cellular automata
- ☐ Probability

Introduction

Can anyone be a genius? I believe that they can. And certainly, if you've taken the trouble to pick up this book, I believe that *you* can.

Perhaps you are shaking your head in disbelief now. After all, few people in history have the distinction of being called geniuses. It's unlikely that you solved one of the Millennium Prize Problems last week. And maybe you don't even think you're good at maths at all.

But what *is* a genius? The mathematicians (and scientists, philosophers and artists) in this book who are called geniuses, and who have produced some truly profound ideas, would not attribute their success to some kind of innate talent, but to hard work and luck. 'Nothing comes from nothing', says Shakespeare's King Lear. Ideas do not spring from a void, but are created through reading other people's ideas, discussing them, playing with them, making mistakes and persevering. Being wrong is as important to genius as being right, because it is only through mistakes that we find solutions to problems.

Just as, with time and training, most of us can become good writers, singers or runners, most of us are capable of excelling at mathematics too. *Get Smart: Maths* won't provide a shortcut to solving the Riemann hypothesis and winning a million dollars, but it will provide an introduction to 50 different fundamental concepts in mathematics, which I hope will inspire you to delve deeper, read further, make conjectures, start discussions and keep exploring. Maybe you will be blown away by the idea that infinity comes in different sizes, or maybe you will be intrigued by the thought of creating a new tiling pattern. Maybe you will search for a new prime number, or maybe you will try to get rich by beating the casinos.

This book need not be read sequentially, but most chapters will make more sense if they are read in order with the other chapters in their section. For example, the chapter on the Poincaré conjecture will make better sense if read after those on Topology and Identifying Shapes. And while the ideas may be unfamiliar, the structure of each chapter is the same. Five 'Are You a Genius?' questions should help to gauge your current understanding of a topic as well as encouraging you to read the chapter to find out the answers! (The answers are on the last page of each chapter, but see if you can figure out the answers yourself after reading the main text.) 'Ten Things a Genius Knows' will lead you through key aspects of the topic, discussing its main ideas, their history, our latest understanding and any unsolved problems. 'Talk like a Genius' provides interesting talking points for dinner parties, while the 'Bluffer's Summary' should allow you to pretend you know what you're talking about, even if the details have made little sense!

Anyone can be a genius: all you need is an idea that excites you. I hope you find excitement in the ideas contained within this book, and I hope you will let me know if it sets you on a path of exploration that leads you to the next big idea.

Julia Collins

Number systems

'God created the integers and all else is the work of man.'

LEOPOLD KRONECKER

Civilizations around the world have developed different systems for representing numbers, from tally marks to Roman numerals, and from Babylonian marks in wet clay to our modern Hindu-Arabic numerals. While all of these systems allowed humans to investigate numbers as abstract concepts, it was the invention of a place-value system that was the truly genius idea. It is through a place-value system that we can work with infinite decimals as well as whole numbers and fractions, giving us a way of expressing every conceivable number. Without it, the modern world, with all its technology, science and economics, could not exist. And yet the idea is not without its own paradoxes and difficulties.

Ever struggled with long division or infinite decimals? Don't worry – you've been grappling with what is arguably mankind's greatest invention.

1 In Roman numerals, CXXV × VII = DCCCLXXV.

TRUE / FALSE

2 The Babylonians had no symbol for zero.

TRUE / FALSE

3 In a base 20 system, we would write $1/4$ as 0.5.

TRUE / FALSE

4 The number 0.999 . . ., with infinitely many 9s, is equal to 1.

TRUE / FALSE

5 Our decimal system can represent numbers that other positional systems cannot.

TRUE / FALSE

TEN THINGS A GENIUS KNOWS

1 How numerals were invented
Are numbers invented or discovered? Philosophers may never agree on the answer to this, but in either case the concept of a number is arguably mankind's most genius idea. In English, we have many number words that are related to particular objects: we talk about a *brace* of pheasants, a *quartet* of musicians, a *dozen* eggs, a *score* of miles, a *grand* (of money). But inventing a word and a symbol for the abstract concept of 'three', as it applies to any object, was a great step of sophistication. Sumerians, in about 3100 BCE, are thought to have been the first to invent written numerals. They used clay tablets to record trade and realized that having a symbol for 'ten' and a symbol for 'sheep' was more efficient than using the sheep symbol ten times.

2 Different ways to design numerals
When designing symbols for numbers, which numbers should we give symbols to? If every number had its own symbol, there would be too many to remember, but if very few had symbols, then writing out large numbers would take lots of space. Imagine writing out 2018 using only tally marks! Many early civilizations, such as the ancient Egyptians, Romans and Greeks, used an additive numerical system. In this method, symbols are created for special numbers (usually 1, 10, 100, 1000, etc., and sometimes also 5, 50, 500) and they are added up to create the required number. So, in the Roman system, X was 10, V was 5 and I was 1, so XXVIII = 10 + 10 + 5 + 1 + 1 + 1 = 28. In this system addition is easy, but multiplication is almost impossible without the use of a tool like an abacus.

3 How place value works
Another method of writing numbers is using a *positional* system, otherwise known as place value. In this system, the position of a numeral within a number determines its value. For example, 123 is different from 321 in our decimal system. This is because the second column from the right multiplies a digit by 10, the third column by 100, the fourth column by 1000, and so on, in increasing powers of 10. This special value of 10 is called the base. It determines how many symbols are needed in the system. Since we use base 10, we need ten symbols:

0, 1, 2, ... 9. Binary numbers use base 2 and need only two symbols: 0 and 1, with the columns representing increasing powers of 2. The Babylonians used base 60, combined with an additive notation for numbers of 59 or less.

4 The importance of zero
A positional system has many advantages over an additive one: few symbols are needed, new symbols are not needed for larger numbers, and multiplication is much easier. But for a positional system to work effectively, it is important to be able to indicate when a column is empty, otherwise the numbers 110, 101 or 11 would be indistinguishable from one another. The Babylonians used a large base, so context was normally sufficient to determine their numbers without the need for an extra symbol, but later they did invent a symbol for zero, as did the Mayans (who used base 20). These symbols for zero were not seen as numbers in themselves, but as placeholders for empty columns. The true use of zero as a number that could be added to, or multiplied with, other numbers was developed in India in the seventh century by the mathematician Brahmagupta, who was also the first to provide rules on how to use negative numbers in mathematics.

5 Using place value to write fractions
A positional system provides an easy way to represent numbers less than 1. Just as moving one column to the left multiplies a number by the base, moving one column to the right divides a number by the base. In our decimal system, the column to the right of the units is worth $1/10$, the second column to the right is worth $1/100$ and so on, so 0.25 means $2 \times (1/10) + 5 \times (1/100) = 25/100 = 1/4$. In the Babylonian base 60 system, $1/2$ is $30/60$ so would be written 0;30, while $5/8$ is $37 \times (1/60) + 30 \times (1/3600)$ so would be written 0;37,30. (Here the semicolon indicates the first column after the units, to distinguish it from the decimal point.)

6 What makes a good base
Not all fractions are straightforward to write in base 10. For example, we cannot write the fraction $1/3$ except as 0.333..., using an infinite number of decimal places. As 10 is 2×5, any number that is not a multiple of only 2s and 5s will have this same problem. In this regard, the base 60 system of the

Babylonians is better than ours, as there are many more fractions that can be written as finite numbers. For example, $1/3$ is 0;20 and $1/6$ is 0;10. But no matter what base is chosen, there will always be fractions that need an infinite number of decimal places to be represented. The Babylonians were not able to write down $1/7$ in their system, and instead used the closest finite approximation that they could find. Modern computers will similarly develop rounding errors whenever infinite decimals are involved.

7. The misconceptions about infinite decimals

Infinite decimals cause trouble both for computers and for our intuition. It is a difficult thing to think of the number 0.333 . . . as genuinely having an infinite number of 3s, rather than being a process that is getting closer and closer to the value of $1/3$. Even if we believe that $1/3 = 0.333$ . . ., multiplying both sides by 3 gives us the unpalatable result that $1 = 0.999$ . . .. Most people, while happy to believe that 0.333 . . . is really $1/3$, find it hard to believe that 0.999 . . . is really equal to 1, thinking of it instead as a number that is just slightly below 1. Mathematically, there is no problem with these equalities, but they show why place value can be hard to teach and understand.

8. That numbers are not always unique

The fact that 1 is equal to 0.999 . . . causes a problem, not only because it challenges our intuition, but because it means that some numbers can be written in more than one way. Similarly, we can show that 0.25 also equals 0.24999 . . ., and 5.341 equals 5.340999 . . .. Indeed, any finite decimal has a second way of being written with infinitely many 9s. This is not an issue specific to our decimal system: a positional system with any base will have this property.

9. What real numbers are

The collection of all numbers that have finitely many digits to the left of the decimal point, and infinitely many after it, are called the real numbers. They encompass all the integers (whole numbers, positive and negative), rational numbers (numbers that can be written as one integer divided by another), and irrational numbers (numbers that are not rational). These numbers are usually visualized as lying on a number line, with numbers getting larger to the right and smaller to the left. Zero is in the centre, marking the transition from positive to negative numbers. A positional system using any base greater than, or equal to, 2 is capable of representing all real numbers.

10. Why real numbers are the basis of mathematics

The construction of the real numbers is considered to be the foundation of all modern mathematics, and this would not have been possible without a place-value system. Throughout history, mathematics has advanced by people 'discovering' new numbers, such as zero, negative numbers, and irrational numbers, but we can show now that the real numbers are complete – there are no gaps and no numbers we have not yet found. Real numbers allow the study of continuous processes, including concepts such as limits and calculus, and hence modern science could not exist without their invention.

TALK LIKE A GENIUS

6 Vestiges of the Babylonian base 60 system can be seen in the way that we have 60 seconds in a minute, 60 minutes in an hour, and 360 degrees in a circle.9

6 During the French Revolution, decimal time was introduced and used for a number of years. The day was divided into ten hours; each hour had 100 minutes and each minute had 100 seconds. A month was divided into three weeks of ten days each. The new system did not catch on, and attempts to decimalize time were finally abandoned by 1900. The Chinese, however, have had many successful decimal calendar systems throughout their history. 9

6 In modern society we are often conflicted over whether the digit 0 is an actual number or a placeholder for an empty column. This can be seen on the computer keyboard, where 0 appears after the 9 (whereas, as a numeral, it should come before the 1), and in our calendar system, where there is simply no year 0 at all. The word "zero" traces its origins back to the Arabic *sifr*, meaning "empty".9

WERE YOU A GENIUS?

1 TRUE – In Roman numerals, D = 500, C = 100, L = 50, X = 10, V = 5 and I = 1. Therefore, this calculation states that $125 \times 7 = 875$, which is true.

2 FALSE – This was true in the early days of their civilization, but the Babylonians later developed a symbol for when a column was to be left empty.

3 TRUE – The column immediately to the right of the decimal point is worth $1/20$, so 0.5 represents the number $5/20 = 1/4$.

4 TRUE – One proof asks: what is $1 - 0.999\ldots$? The answer is smaller than any real number and so it must be zero.

5 FALSE – All positional systems are capable of representing all real numbers.

THE BLUFFER'S SUMMARY

The invention of a place-value system, where a digit has different values depending on where it appears within a number, is what gave humans the ability to write down every conceivable quantity, not just whole numbers.

Binary numbers

'There are 10 types of people in the world: those who understand binary, and those who don't.'

UNKNOWN

The simplest form of counting is also the one that is the most integral to our modern, technology-based society, being the language in which computers are programmed. Understanding the limitations of binary is key to avoiding errors, even on something as simple as a spreadsheet. Binary numbers are used in other ingenious ways to catch and correct errors in data transmission, making it possible to send messages across the world and even across the solar system.

How is it that just two little symbols – 0 and 1 – have become the language of all technology?

1 The binary number 10101 is equal to 21.

TRUE / FALSE

2 Some whole numbers can be written in a number of different ways using binary.

TRUE / FALSE

3 A byte in a computer is a single binary digit, represented by an on/off switch.

TRUE / FALSE

4 Using the decimal number 0.1 in any computer calculation will result in a rounding error.

TRUE / FALSE

5 Digits can be added to binary numbers that can detect and correct errors.

TRUE / FALSE

TEN THINGS A GENIUS KNOWS

1 What binary numbers are
In binary there are only two symbols: 0 and 1. To represent a number using only these two symbols, we use a positional system with base 2, so the columns represent the 1s, the 2s, the 4s, the 8s, and so on, with each column worth twice that of the one to its right. (This is the same idea as in our decimal notation, where each column is worth ten times that of the one on its right.) We call the numbers 1, 2, 4, 8 . . ., powers of 2 because they are repeated multiples of 2, and we write them as $2^0, 2^1, 2^2 \ldots$, with the superscripts indicating how many multiples of 2 create the number. The numbers 1, 2, 3, 4, 5, 6, 7 in binary are 1, 10, 11, 100, 101, 110, 111. The binary number 10011 is equal to the decimal number $(1 \times 16) + (0 \times 8) + (0 \times 4) + (1 \times 2) + (1 \times 1) = 19$. To avoid ambiguity between decimal numbers and binary numbers, the notation of $(.)_2$ can be used for binary numbers, so $(101)_2$ means five, not one hundred and one.

2^4	2^3	2^2	2^1	2^0
16	8	4	2	1
1	0	0	1	1
0	0	1	0	1

$00101 = 4 + 1 = 5$ $10011 = 16 + 2 + 1 = 19$

2 How to convert a decimal number into binary
Every decimal number can be written in a unique way using binary. One way to convert a decimal number into binary is to repeatedly halve the number, keeping track of the remainders at each stage.

To convert 27 into binary we do the following: halve 27 to get 13 with remainder 1, then halve 13 to get 6 with remainder 1, then halve 6 to get 3 with remainder 0, then halve 3 to get 1 with remainder 1, then halve 1 to get 0 with remainder 1.

$27 \div 2 = 13$	rem 1	1
$13 \div 2 = 6$	rem 1	1
$6 \div 2 = 3$	rem 0	0
$3 \div 2 = 1$	rem 1	1
$1 \div 2 = 0$	rem 1	1

The sequence of remainders were 1, 1, 0, 1, 1, so 27 is written in binary as $(11011)_2$. Indeed, $27 = 2^4 + 2^3 + 2^1 + 2^0 = 16 + 8 + 2 + 1$, so the answer is correct.

3 The invention of binary numbers
Many ancient civilizations around the world and throughout history have represented numbers using only two symbols. They include the Chinese, Polynesians, Indians and indigenous Australians. However, German mathematician Gottfried Leibniz is credited with inventing our modern binary number system in 1679. Designing it initially as a system for encoding logical statements, Leibniz realized that it was the perfect system for building a mechanical calculator that could use 'moving balls to represent binary digits'. Though he never built such a device, the idea became the basis for modern computers.

4 How computers use binary
The 0 and 1 of binary are represented in a computer by the on/off of a switch. In this way a bank of eight switches can represent all the numbers from 0 to 255. A single binary digit in a computer is called a 'bit', while eight bits is called a 'byte'. A byte is usually the smallest unit of memory in a computer; this is because it is both a convenient power of 2 and also the number of bits needed to encode a single character of text. For example, the ASCII system uses bytes to encode uppercase and lowercase letters, the numerals 0–9, basic punctuation symbols, a space, and commands such as 'return' and 'backspace'.

5 How to add in binary
A circuit that adds binary numbers, called a binary adder, is one of the most basic and important components of any computer. The simplest version, called a half-adder, adds up two single binary digits. If both digits are 0, the answer is 0. If one digit is a 0 and the other is a 1, then the answer is 1. If both digits are 1, then the answer is 0 with a 1 carried over to the next column. (This is similar to adding 5 + 5 in decimal and getting 0 with a 1 carried over to the 10s column.) For example, if we add 101 to 011 then we first add the rightmost digits to get $1 + 1 = 0$ with a carry of 1. Next, we add the middle two digits plus the carry to get $0 + 1 + 1 = 0$ with a carry of 1. Finally, we add the first two digits plus the carry to get $1 + 0 + 1 = 0$ with a carry of 1. So the answer is 1000. (In decimal, this calculation is saying that 5 + 3 = 8.)

6. The limitations of binary

We can represent fractions in binary, as we can in any positional system. The column to the right of the decimal place represents $1/2$, the next column to the right represents $1/4$, the next $1/8$ and so on. If we write the number $(0.11)_2$, this is $1/2 + 1/4 = 3/4$. But only fractions whose denominator is a power of 2 can be written down using finitely many digits. The number $1/3$ in binary is $0.010101\ldots$ with infinitely many 01s, and $1/10$ is $0.000110011\ldots$ with infinitely many 0011s. Since computers only have a fixed amount of memory, they have to round off these numbers, and this is called a 'floating point error'. Working in base 10, and continually converting to binary, is a particular problem, and can be seen in a simple spreadsheet. Start with a cell containing 1 and successively subtract 0.1 until you reach 0. Due to rounding errors, the computer will believe this value to be slightly bigger than 0; more specifically, about 1.38×10^{-16}.

7. How to use binary to check for errors

When information is sent, there is always a possibility for error. A signal sent to the Mars rover can have a digit corrupted as it traverses the solar system, making the rover fall off a cliff by turning left instead of right. To detect these possible errors, a 'parity bit' is added to the end of a binary number. If the number of 1s in the original message is an odd number, then the parity bit is 1, and if the number of 1s in the original message is even, then the parity bit is 0. A machine receiving the message 1101011, where the last digit is a parity bit, will know that the message is corrupt because there are four 1s in the original message, but the parity bit is not 0. This means that one of the digits has changed between the sending and receiving of the message.

8. How parity bits can correct errors

A single parity bit can detect that a message has an error, but it cannot say which digit is wrong. A 'Hamming code' adds a number of parity bits to a message, allowing errors not only to be detected, but also corrected. Again, it uses binary numbers to achieve this feat. The digits that are in positions that are powers of 2 (that is, the 1st, 2nd, 4th, 8th positions) are the parity bits. The parity bit in position one checks all digits whose positions are odd (that is, the positions whose binary representation has a 1 in the units column). The parity bit in position two checks all digits whose binary representation has a 1 in the 2s column, and so on. This means that if, for example, the parity bits in positions one, four and eight are wrong, then the digit in position $1 + 4 + 8 = 13$ is the one that has been corrupted.

9. The most efficient way to correct errors

There are other ways in which we could detect and correct errors in a message. For example, we could transmit each binary digit three times, so that instead of sending 1011, we would send 111 000 111 111. If we received the message 111 001 111 110, we could still reconstruct the original message by taking the majority digit in each set of three, reasoning that it is unlikely that two of the extra check digits would be wrong. The drawback to this method is that the messages being sent are three times longer than the original. The Hamming code described above is the most efficient way of detecting and correcting single-digit errors. For a four-bit message, three extra check digits are required, instead of the eight extra needed for the 'repeat each digit three times' method.

10. How printers use binary to encode secret data

In 2005, it was revealed that almost all brands of colour printer were printing miniscule yellow dots on every page, using binary code to identify the time, date and individual printer used to create a document. The dots are not visible to the naked eye, but appear under blue light using a microscope or magnifying glass. In the example of a Xerox printer (below), each row represents binary digits, while columns represent different data types. The first row and first column are parity bits, confirming the data has printed correctly. The codes for most printers are still yet to be decoded.

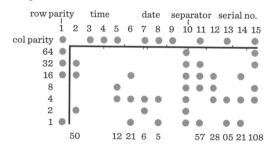

TALK LIKE A GENIUS

❦ Contrary to what most people think, a kilobyte does not always mean 1000 bytes, as it should in the metric system. Since computers work in powers of 2, it is common for programmers to use the word kilobyte to represent $2^{10} = 1024$ bytes. While the difference isn't very much for a kilobyte, the metric/binary differences are noticeable for megabytes, gigabytes and terabytes. To distinguish between the two uses, a lowercase k is used for the metric system (for example, kB) and an uppercase K for the binary system (for example, KB). This is also why phone or electronic device memory rarely corresponds to the advertised number. ❧

❦ Hamming codes can be used for the following party trick. Ask someone to think of a number between 0 and 15. You can ask them seven yes/no questions, and they can lie at most once. Using the idea that four of the questions are checking for data (that is, the four binary digits of the number) and three of the questions are parity bits, you can impress your audience by saying not only the number that was thought of but also which of the answers was the lie. ❧

❦ In June 2017, the FBI were investigating top-secret documents that had been leaked to the press. From creases on the document it was clear that it had been printed and folded. Secret yellow microdots placed by the printer were later used to verify the date and time of the printing, confirming that it was not a forgery. ❧

WERE YOU A GENIUS?

1 TRUE – $10101 = (1 \times 16) + (0 \times 8) + (1 \times 4) + (0 \times 2) + (1 \times 1) = 21$.

2 FALSE – Every whole number has a unique binary representation.

3 FALSE – A binary digit in a computer is called a 'bit' and eight bits is a 'byte'.

4 TRUE – When the number 0.1 is written in binary, it has infinitely many binary digits, so memory limits on a computer will always result in rounding errors.

5 TRUE – Parity bits added to numbers in clever ways can detect if the number is correct and, if it is not, can tell where the error is.

THE BLUFFER'S SUMMARY

A binary system represents all numbers using just the symbols 0 and 1, and is the basis for how all computers work.

Irrational numbers

'There is geometry in the humming of the strings, there is music in the spacing of the spheres.'

PYTHAGORAS

Pythagoras believed that everything in the world could be explained by whole numbers and their ratios, from the harmonies in music to the very form of our solar system. When his disciple, Hippasus, demonstrated that there existed numbers that could not be expressed as fractions, Pythagoras had him drowned. These numbers became known as 'irrational', invoking a continued mistrust of their character. But it was not until the 19th century that mathematicians discovered an even darker and more enigmatic side to irrational numbers, which, today, many are still trying to understand.

The overwhelming majority of numbers have no pattern in their decimal expressions, creating a constant struggle for the mathematicians (and computers) trying to make sense of them.

1 The digits of π will eventually repeat, though they may take a long time to do so.

TRUE / FALSE

2 The sum of two irrational numbers is always an irrational number.

TRUE / FALSE

3 The imaginary number $i = \sqrt{-1}$ is transcendental.

TRUE / FALSE

4 There are only finitely many transcendental numbers.

TRUE / FALSE

5 Some numbers cannot be described using any finite formula or algorithm.

TRUE / FALSE

TEN THINGS A GENIUS KNOWS

1 What an irrational number is
An irrational number is one that cannot be written as a ratio of two whole numbers. The first such number to be discovered was the square root of 2. This number clearly exists: it is the length of the hypotenuse in a right-angled triangle with two sides of length 1. But any attempt to show that it can be written as a ratio of two whole numbers will lead to a contradiction. Such an argument was hinted at by Aristotle and first proved in full by Euclid in his *Elements*. In fact, the square root of any number that is not a perfect square (like 4 or 25) is easily proved to be irrational. A more interesting example of an irrational number is π, which was only proved irrational in 1761.

2 The problem with irrational numbers
Irrational numbers are difficult to deal with because they can never be written down exactly. A rational number (one that can be written as a ratio of whole numbers) will always have a repeating decimal. Some terminate completely, such as $1/2 = 0.5$; some repeat quickly, such as $1/3 = 0.3333\ldots$; and others repeat on a longer scale, such as $1/19$, which repeats every 18 digits. Conversely, every repeating decimal can be written as an integer fraction. So, an irrational number must have a decimal expansion that never repeats itself. (This is true in every base, not just base 10.) This can make computing irrational numbers very difficult. Some people dedicate their lives to finding the next digit of π! It also means that, when irrational numbers are used in computations, there will be rounding errors because it is impossible for computers to represent such numbers exactly.

3 How to think about numbers algebraically
One way to get around the difficulty of writing down irrational numbers is not to think of what they are but what they *do*. The Egyptian mathematician Abū Kāmil in the ninth century CE was the first to think about numbers as solutions of equations. He thought of $\sqrt{2}$ as the number that solves the equation $x^2 = 2$. That is, $\sqrt{2}$ is the number whose square is 2. This makes it easy to manipulate the number algebraically without worrying about what its exact value is. Another example is the golden ratio φ whose value is $(1+\sqrt{5})/2$, and which can be thought of as the number that solves the equation $x^2 - x - 1 = 0$.

4 What algebraic numbers are
The imaginary number $i = \sqrt{-1}$ does not fit on our familiar number line and does not make much sense in terms of anything that exists in the real world. But using Abū Kāmil's idea, we can represent it as the solution of $x^2 = -1$ and work with it algebraically. It is neither rational nor irrational, but it is an *algebraic* number: it is the solution of a polynomial equation with integer coefficients. A polynomial is the sum of powers of a variable x (that is, x, x^2, x^3 etc.) and its coefficients are the numbers that multiply these powers. So the expression $2x^5 + 52x^3 - 7 = 0$ is an integer polynomial equation, but $x^3 + \sqrt{3}\,x = 0$ is not because $\sqrt{3}$ is not an integer.

5 Whether all numbers are algebraic
Mathematicians wondered whether all numbers were algebraic and if Pythagoras was right after all, in asserting that whole numbers were the key to understanding the universe. Then, in 1884, Joseph Liouville constructed a number that he could prove was not the solution to any polynomial equation involving whole numbers. This number is $0.110010000000000000000001000\ldots$ – it has a 1 in the decimal places that are factorial numbers. (These are numbers that are products of all smaller numbers, so $1! = 1$, $2! = 1 \times 2$, $3! = 1 \times 2 \times 3$, and so on.)

6 What a transcendental number is
A transcendental number is a number that is not algebraic. Liouville had constructed the first example of such a number 'artificially', but in 1873 Charles Hermite showed that e, the mathematical constant that arises naturally in calculus, is also transcendental. Within ten years of Hermite's proof, Ferdinand von Lindemann proved that π was transcendental. So-called 'continued fractions', written in the style shown here, offer another way of constructing many transcendental numbers. We can find decimal approximations for such numbers by truncating the continued fraction further and further

down. In the example here, the first approximation is 1, while the second is $1 + \frac{1}{2} = 1.5$. The third approximation is $1 + 1/(2 + \frac{1}{3}) = 1 + \frac{3}{7}$ or 1.42857, and so on. The further down we truncate, the closer the value of the decimal will come to the true value of the continued fraction.

$$1+ \cfrac{1}{2+ \cfrac{1}{3+ \cfrac{1}{4+ \cfrac{1}{5+ \cfrac{1}{6+ \ddots}}}}}$$

7 What we know (and don't know) about transcendental numbers

Despite Hermite and von Lindemann's achievements over 100 years ago, mathematicians have still not been able to prove some basic facts about transcendental numbers. They do not know whether or not π^π is transcendental, nor whether $\pi^{\sqrt{2}}$ is. They can show that at least one of $\pi + e$ and πe must be transcendental, but they cannot prove which one. Surprisingly, they do know that πe^π is transcendental. The problem of deciding whether a number is transcendental was considered so important that it became part of Hilbert's list of 23 problems posed in the famous 1900 Congress of Mathematicians. Hilbert's 7th problem asked for a proof of whether a^b was always transcendental if a was algebraic (and not 0 or 1) and b was irrational. Gelfond and Schneider provided the proof in 1935.

8 How many transcendental numbers exist?

Given how few transcendental numbers are known, it would be easy to believe that they are a rare breed of numbers, occasionally encountered, but nothing to concern the average person. In fact, not only are there an infinite number of them, but there are an uncountable infinity of them. (See page 53). *Most* numbers are transcendental. If you threw a dart at a number line, the dart would almost certainly land on a transcendental number.

9 Methods for computing transcendental numbers

Transcendental numbers cannot be written in terms of simple equations, but this does not mean that they cannot be computed to any desired precision by a computer. There are often other ways to express such numbers, such as continued fractions (as in the example above), or infinite series. The number e, which is incredibly important in science, is straightforward for a computer to calculate, using formulae such as:

$$\sum_{n=0}^{\infty} \frac{1}{n!} = \frac{1}{0!} + \frac{1}{1!} + \frac{1}{2!} + \frac{1}{3!} + \dots$$

– that is, adding the fractions with factorial numbers in the denominators. It is an infinite sum, so of course a computer cannot calculate it entirely, but it can find the answer to any finite precision that is needed for a calculation.

10 What uncomputable numbers are

The final blow to Pythagoras' vision of the universe came in the 20th century, when men such as Alan Turing and Marvin Minsky showed that most numbers are not only transcendental, but are also uncomputable. A number is uncomputable if there is no finite algorithm or formula that can compute its digits to any required degree of accuracy. The existence of such numbers is unlikely to impact our everyday lives, although it does have a deep part to play in the logic underpinning mathematics and the question of whether computer programs can finish running in a finite amount of time.

TALK LIKE A GENIUS

❦ The golden ratio is sometimes called the 'most irrational' of numbers because it is the number that is most difficult to accurately approximate by fractions. This is because it can be written as a continued fraction with 1s all the way down, and truncations of this continued fraction produce a sequence of fractions that take a very long time to get close to the true value of the golden ratio. The larger the numbers within the continued fraction, the better the approximations will be. ❧

❦ The number π is currently known to over 22 million million decimal places, with the most recent record breaker being in 2016. The computation took over 100 days. A perhaps more impressive record breaker is William Shanks, who in 1873 computed 527 digits entirely by hand. ❧

❦ The proof that π is transcendental resolved once and for all that it was impossible to "square the circle"; that is, to find a square with the same area as a given circle using only a ruler and a compass. This had been an open question since Babylonian times. In 1894, a man named Edward Goodwin was so convinced that he had a method to square the circle that he proposed a bill in the Indiana General Assembly that his results should be taught in school. This would have meant enshrining in law that π was exactly 3.2. Thankfully, the bill did not become law, although it did pass its first reading. ❧

WERE YOU A GENIUS?

1 FALSE – π is an irrational number, and these numbers can never be represented by a repeating decimal number.

2 FALSE – For example $\sqrt{2} + (1 - \sqrt{2}) = 1$.

3 FALSE – It is a solution of the equation $x^2 + 1 = 0$, so it is an algebraic number.

4 FALSE – There are uncountably infinitely many transcendental numbers.

5 TRUE – Such numbers are called uncomputable numbers.

THE BLUFFER'S SUMMARY

Most numbers are not only irrational but are transcendental, meaning that they cannot be described by finite equations using only whole numbers.

Imaginary numbers

'The imaginary number is a fine and wonderful resource of the human spirit, almost an amphibian between being and not being.'

GOTTFRIED WILHELM LEIBNIZ

Throughout history, people have struggled with different types of number, from negative numbers to zero to infinite decimals, but it is square roots of negative numbers that have stretched people's imagination the most. But, imaginary or not, these numbers uncannily explain the world in which we live, making them an indispensable tool for any budding genius.

Modern physics explains the world using numbers that do not even exist. This paradox is as hard to imagine as imaginary numbers themselves.

1 Imaginary numbers (other than zero) do not exist anywhere on the regular number line.
TRUE / FALSE

2 The square root of any negative number can be written as a multiple of i, the square root of –1.
TRUE / FALSE

3 It is impossible to divide a real number by an imaginary number.
TRUE / FALSE

4 It is impossible to square root an imaginary number without inventing another new kind of number to deal with the answer.
TRUE / FALSE

5 Quantum physics needs both real and imaginary numbers so that it can keep track of a particle's position and momentum at the same time.
TRUE / FALSE

1 How to think about the square root of a negative number

What is –1? This number does not really 'exist', in the sense that we cannot have –1 sheep in a field. But we can deal with it mathematically by thinking about what it does: it is the number that, when added to 1, gives zero. (The farmer with –1 sheep is in debt, so that once they give a sheep to someone, they will have no sheep.) A similar thing happens with the square root of 2 (the number that multiplies by itself to give 2): we cannot truly make sense of it numerically because it is an infinite decimal that never repeats. But we know it is a number that, when squared, gives an answer of 2. When mathematicians came across the idea of the square root of –1, they could not make sense of it, because there is no number on the number line that can multiply by itself to give a negative number. But they gave it a symbol, i (for imaginary), and were able to use it in calculations by knowing that $i^2 = -1$.

2 What a complex number is

We invent a special symbol, i, for $\sqrt{-1}$, but what happens for the square roots of all the other negative numbers? Thankfully, we do not need symbols for all of them, because the square root of any negative number can be written as a multiple of i. For example, $\sqrt{-9} = \sqrt{(9 \times -1)} = \sqrt{9}\sqrt{-1} = 3i$. The square root of any negative number is called an imaginary number. Imaginary numbers can be added to real numbers, forming complex numbers. A complex number has the form $a + bi$, where a and b are real numbers. The number a is called the real part, while b is called the imaginary part.

3 How to visualize complex numbers

Square roots of negative numbers clearly do not fit anywhere on the number line. To visualize them, we add in a new axis at right angles to the real number line. This picture is called the complex plane and is like a coordinate grid, with real numbers along the horizontal axis and imaginary numbers up the vertical axis. A complex

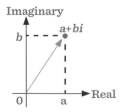

number $a + bi$ has coordinates (a,b) in the picture. For example, $2 + 3i$ has coordinates $(2,3)$, while $5 – 2i$ has coordinates $(5,–2)$.

4 How to do arithmetic with complex numbers

Complex numbers can be added, subtracted, multiplied and divided, just like any other numbers, but a little algebra is needed to figure out how to do this properly. Addition and subtraction are straightforward: add the real parts and the imaginary parts of the complex numbers separately. For example, $(2 + 3i) + (5 – 2i) = (2 + 5) + (3 – 2)i = 7 + i$. Multiplication requires us to remember how to multiply out brackets, and that $i^2 = -1$. So, $(2 + 3i) \times (5 – 2i) = (2 \times 5) + (2 \times -2i) + (3i \times 5) + (3i \times -2i) = 10 – 4i + 15i – 6i^2 = 10 + 11i + 6 = 16 + 11i$. Division is the most complicated operation and is done using a clever trick involving the complex conjugate of the number. If a complex number like $2 + 3i$ is multiplied by the same number but with the imaginary part made negative $(2 – 3i)$, then the answer is always a real number. Dividing by a complex number is the same as multiplying by its complex conjugate and then dividing by this real number.

5 The relationship between complex numbers and trigonometry

One reason that complex numbers are so important in science is that they are very closely related to the trigonometric functions of sine and cosine. They are, in some sense, a bridge between the mathematical areas of algebra and geometry. The position of a complex number in the complex plane can be given by its real and imaginary coordinates, but it can also be given by finding its distance, r, from the origin and the angle, θ, it makes from the positive real axis. Using some basic trigonometry, we can write a complex number $a + bi$ as $r(\cos\theta + i\sin\theta)$. Any area of physics based on wave functions, such as electromagnetism, requires trigonometry, and so can be modelled by complex numbers.

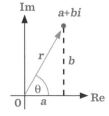

6 How to do rotations using complex numbers

The relationship between complex numbers and trigonometry goes deeper than just being able to use sine and cosine to represent complex numbers. We can use complex numbers to do transformations of images, such as scaling and rotation. Imagine a graphic designer who has an image that they would like to rotate 90 degrees anticlockwise and scale to be twice as big as before. Traditionally, this would be done using a 2 × 2 matrix, which involves four different multiplications and two additions for each point in the image – a process that becomes time-consuming when the image is large. But the same result can be achieved by multiplying each point by the complex number $2i$. It turns out that whatever combination of scaling and rotation the designer wants to achieve, there is a single complex number that can do this in one multiplication.

7 How complex numbers 'complete' the real numbers

Without imaginary numbers, not every quadratic equation has a solution. For example, the equation $x^2 + 1 = 0$ has no solutions if x is restricted to being a real number. Allowing x to be a complex number, it turns out that every quadratic equation has a solution. When we look at cubic equations – for example, $x^3 + x + 1$, we may wonder whether we will need to invent yet more numbers in order to ensure that we will always get a solution. Amazingly, we do not. No matter how big our equations get, allowing x to be a complex number means that we will always find a solution. This is called the fundamental theorem of algebra and, for mathematicians, is one of the most important properties of complex numbers. The complex numbers are said to 'complete' the real numbers because they can solve any polynomial equation involving real numbers.

8 Why complex numbers are used in quantum physics

Imaginary numbers are at the heart of quantum physics, from Schrödinger's equation to Heisenberg's uncertainty principle. Why is it not possible to write these equations using only real numbers? One reason is that quantum physics needs 'two-dimensional' numbers to encode basic principles. A quantum system is modelled using a wave equation that makes predictions about the probability of particles being in particular places at particular times. If this equation is created using only real numbers then it predicts that the momentum of the particles is always zero, which is not true. But if the equation is formulated using complex numbers, then it can keep track of both the position and momentum of the particles at the same time. The special arithmetic of the complex numbers is also important here: it is things such as the use of complex conjugates that make the predictions of the wave equation come out correctly.

9 How to make four-dimensional complex numbers

Complex numbers are indispensable tools to scientists when modelling systems that require two coordinates, such as electromagnetic waves or quantum systems. Mathematicians wondered whether there was a way of expanding the notion of complex numbers to model three-dimensional (3-D) systems. It turns out that there isn't, but there *is* a way of expanding complex numbers to a 4-D system. These numbers are called the quaternions, and they have three different types of imaginary number: i, j and k. The quaternions have the strange property of being non-commutative. That is, multiplying two quaternions in a different order will usually give a different answer.

10 Where quaternions are useful

A quaternion is written as $a + bi + cj + dk$ where a, b, c and d are real numbers. Just as multiplication by complex numbers allows us to do rotations in 2-D, multiplication by quaternions allows us to do rotations in 3-D, and it turns out to be far more efficient than any other method. One alternative is to use a rotation matrix, but this is a 3 × 3 object so requires nine numbers to specify it, compared with four for the quaternion. Rounding errors obtained by performing many rotations in a row are also minimized by using quaternions instead of matrices, and it is much easier to do a 'smooth rotation', such as camera panning, in a computer game.

TALK LIKE A GENIUS

❦ In the 15th century, Italian mathematicians solved disputes by having duels – albeit with equations rather than swords. One mathematician, Tartaglia, accused a rival, Cardano, of stealing his formula for cubic equations, and so he duelled with Cardano's student Ferrari. Cardano *had* taken Tartaglia's formula, but he and Ferrari had adapted it to solve a wider class of problems, including the novel idea of using square roots of negative numbers to get solutions. Tartaglia lost the duel, lost his job and died in poverty. ❧

❦ The formula for quaternion multiplication came to Sir William Rowan Hamilton as he was walking along the Royal Canal in Dublin. He was so excited by his discovery that he used his penknife to carve the equations into the side of Broom Bridge. The original markings have since been eroded, but a plaque is now there to mark the inscription. ❧

❦ Lord Kelvin claimed that quaternions were "an unmixed evil to those who have touched them in any way", and they were largely ignored by mathematicians and scientists until the late 20th century. Today, they are indispensable in computer graphics, flight dynamics and for satellite navigation. ❧

WERE YOU A GENIUS?

1 TRUE – Imaginary numbers sit on their own number line, which is at right angles to the regular one.

2 TRUE – For example, $\sqrt{-4} = \sqrt{(4 \times -1)} = \sqrt{4}\sqrt{-1} = 2i$.

3 FALSE – There is a method for dividing real numbers by imaginary numbers, giving another imaginary number as the result.

4 FALSE – The square root of any imaginary number is a complex number – that is, a sum of real and imaginary numbers, so no new types of number need to be created.

5 TRUE – If only real numbers are used, the wave equation in quantum physics predicts that all particles have a momentum of zero.

THE BLUFFER'S SUMMARY

An imaginary number is the square root of a negative number and, despite being very hard to imagine, is used to model quantum mechanics, electrical signals and motion in computer games.

Fermat's Last Theorem

'I have discovered a truly remarkable proof of this theorem, which this margin is too small to contain.'

PIERRE DE FERMAT

This tantalizing quote led to 350 years of frustration for mathematicians, who searched in vain for Fermat's proof of his simple conjecture: that two cubes could never sum to make another cube, nor two fourth powers sum to make another fourth power, and so on. Fermat made many other conjectures, all of which were quickly checked by other people, leaving this one as the 'last theorem' to be solved. The eventual proof, in 1995 by Andrew Wiles, was truly a work of genius, bringing together all the major ideas in 20th-century number theory and spawning many new ideas.

Equations using only whole numbers may seem the simplest of all equations, but they turn out to be the most complicated, often keeping mathematicians puzzled for hundreds of years at a time.

1 There are only finitely many right-angled triangles where all the side lengths are whole numbers.

TRUE / FALSE

2 Fermat had no evidence for the truth of his conjecture.

TRUE / FALSE

3 It is possible for a number to be simultaneously both a square and a cube.

TRUE / FALSE

4 Every whole number bigger than 2 is either a multiple of an odd prime number or a multiple of 4.

TRUE / FALSE

5 An obscure conjecture relating elliptic curves to modular forms turned out to be the key to proving Fermat's Last Theorem.

TRUE / FALSE

TEN THINGS A GENIUS KNOWS

1 How many integer-sided right-angled triangles there are

Is it possible to add two square numbers to get another square number? Yes: for example, $3^2 + 4^2 = 5^2$ ($9 + 16 = 25$). Another example is $5^2 + 12^2 = 13^2$. In fact, the ancient Greek mathematician Euclid proved that there were infinitely many solutions to the equation $x^2 + y^2 = z^2$. This result has an interpretation in geometry through Pythagoras' theorem, which says that, in a right-angled triangle, the sum of the squares of the two shorter side lengths gives the square of the hypotenuse. Euclid's result showed that there were infinitely many different right-angled triangles whose side lengths were all whole numbers (integers).

2 What Fermat's Last Theorem says

In 1637, the French mathematician Pierre de Fermat saw Euclid's result and wondered if a similar theorem might be true for higher powers. Is it possible to add two cube numbers to get another cube number? Or to add two fourth powers to get another fourth power? After trying many examples and failing, he conjectured that it was impossible to find such numbers – not just for cubes, but for *any* power. This conjecture became known as Fermat's Last Theorem. Its formal statement is that the equation $x^n + y^n = z^n$ has no positive integer solutions for any value of n greater than 2.

3 What Diophantine equations are

Fermat's equation is an example of a more general object called a Diophantine equation, which seeks integer solutions for polynomial equations with two or more unknowns. For example, we might notice that the square number 9 is one more than the cube number 8, and wonder whether there are any other solutions to the equation $x^2 - y^3 = 1$. Or we might ask whether a number could be the sum of two cubes in two different ways and come up with the Diophantine equation $x^3 + y^3 = z^3 + w^3$. Because of the restriction of only finding integer solutions, Diophantine equations are notoriously difficult to solve. In 1900, David Hilbert challenged mathematicians to find an algorithm that could decide if a general Diophantine equation had solutions, but in 1970 it was finally proved that such an algorithm could not exist.

4 How the first few cases of the conjecture were proven

Although Fermat did not write down a proof of his conjecture, he did give a proof in the particular case when $n = 4$. He did this using a method called 'infinite descent', where he first assumed that such an equation had a solution and then derived a contradiction from this. If there were a positive integer solution to the equation $x^4 + y^4 = z^4$, then Fermat showed how to use this first solution to generate another, smaller solution, and then use that solution to get a smaller one, and so on, creating an infinite sequence of smaller and smaller solutions. Since we cannot keep writing down smaller and smaller positive integers, such a sequence cannot exist, so the first solution could never have existed. A similar approach was used by Leonhard Euler to prove the result for $n = 3$, but mathematicians were unable to make it work for many other higher powers.

5 Why not every case of Fermat's Last Theorem needs to be proved

Although it seems as if proving Fermat's Last Theorem for the cases $n = 3$ and $n = 4$ were just two rungs on an infinite ladder of cases, each of these proofs in turn gives the result for an infinite number of values of n. For example, if a solution existed for the case $n = 6$, so that $A^6 + B^6 = C^6$ for integers A, B and C, then this would mean that $(A^2)^3 + (B^2)^3 = (C^2)^3$, which would be a solution for $n = 3$, but we already know this is impossible. In general, if a solution existed for a composite number pq, then solutions would also exist for p and q. This reasoning means that it is sufficient to prove Fermat's Last Theorem for the odd prime numbers and for the number 4, since any integer greater than 2 is a multiple of these numbers.

6 Why Lamé's wrong solution was interesting

In 1847, Gabriel Lamé presented a supposed proof of Fermat's Last Theorem. His method was to introduce complex numbers in order to factorize Fermat's equation and to try to find a contradiction. If ζ is a complex number with the property that $\zeta^n = 1$, then Fermat's equation can be written as $x^n + y^n = (x + y)(x + \zeta y)(x + \zeta^2 y) \ldots (x + \zeta^{n-1} y)$. However, this line of reasoning was invalid because it implicitly assumed that complex numbers could be factorized uniquely – similarly to how integers can be uniquely factorized into prime numbers. This assumption is wrong when n is a prime bigger than 23. The investigation of this phenomenon led to a whole new field of number theory pioneered by Ernst Kummer, who redefined the notion of a complex prime number and was able to prove Fermat's Last Theorem for infinitely many different primes, though still not all of them.

7 What the Taniyama-Shimura-Weil conjecture is

The next breakthrough in Fermat's Last Theorem came in 1955 when the Japanese mathematicians Goro Shimura and Yutaka Taniyama made a wild conjecture relating two completely different branches of mathematics. An elliptic curve is a special curve drawn in the plane that comes from an equation of the form $y^2 = x^3 + ax + b$, with the additional stipulation that there be no self-intersections or cusps in the curve. A modular form is a mathematical function of complex numbers that is highly symmetric in a number of ways. Taniyama and Shimura believed that it was possible to relate every (rational) elliptic curve to a modular form, but this seemed to be so far-fetched that it was not taken seriously for over ten years, until André Weil found some evidence that the conjecture might be true.

8 The relationship between Taniyama-Shimura-Weil and Fermat

In 1984, the German mathematician Gerhard Frey created new interest in the Taniyama-Shimura-Weil conjecture when he showed how it could potentially be used to prove Fermat's Last Theorem. If Fermat's equation had an integer solution, then this solution could be used to construct an elliptic curve that was so unusual it was highly unlikely to be related to a modular form. So if Fermat's Last Theorem were false, the Taniyama-Shimura-Weil conjecture would be false. Conversely, if Taniyama-Shimura-Weil were true, Fermat's Last Theorem would be true, too. The American mathematician Ken Ribet was able to prove that Frey's intuition was correct, paving the way for a new line of attack on the problem.

9 How Andrew Wiles proved Fermat's Last Theorem

Andrew Wiles is a British mathematician who had harboured a lifelong desire to prove Fermat's Last Theorem. On hearing Frey and Ribet's results he laboured in secret for seven years trying to prove a special case of Taniyama-Shimura-Weil that would imply Fermat's Last Theorem. In 1993, he presented his finished proof to the mathematical community, but when the proof was checked, a gap was found that meant a further year's work was needed to make the argument watertight. The final proof, published in May 1995, brought together a huge number of mathematical ideas, including Galois theory, group theory, Iwasawa theory (related to Kummer's work), algebraic geometry and category theory.

10 Where Fermat's Last Theorem leads next

Although Fermat's Last Theorem is now proved, the ideas that were generated in Wiles's proof have provided new insights and conjectures that are still keeping mathematicians occupied. The full Taniyama-Shimura-Weil conjecture was proved by various people between 1996 and 2001, confirming the unexpected link between elliptic curves and modular forms. This theorem itself has turned out to be a special case of a bigger programme of work called the Langlands conjectures, which is considered to be a 'grand unified theory of mathematics'.

❝Until it was solved, Fermat's Last Theorem was in the *Guinness Book of World Records* as both the longest-standing unsolved mathematical conjecture and also the 'most difficult mathematical problem' due to having the largest number of unsuccessful proofs. The Goldbach Conjecture is the current record-holder for longest-standing problem.❞

❝In 1919, the mathematician G.H. Hardy visited the Indian mathematician Ramanujan in hospital.On arriving he remarked that he had ridden in a taxi with number 1729, which was rather dull. Ramanujan instantly replied that, in fact, the number was very interesting, as it was the smallest number that could be written as the sum of two cubes in two different ways: $1729 = 1^3 + 12^3 = 9^3 + 10^3$. This number solves the Diophantine equation given in point 3, and similar numbers have since been called "taxicab numbers".❞

❝Few mathematicians believe that Fermat really had a proof of his conjecture. Not only was Wiles's proof so incredibly complex as to have been impossible for a 17th-century mathematician to discover, but Fermat often made mistakes in his other conjectures. For example, he believed that $2^{2^n} + 1$ was prime for every n, but it fails already when $n = 5$.❞

WERE YOU A GENIUS?

❚ FALSE – The sides of a right-angled triangle, a, b and c, are related by Pythagoras' theorem $a^2 + b^2 = c^2$ and this has infinitely many whole number solutions.

❷ FALSE – Fermat proved that his conjecture was true in the case of fourth powers.

❸ TRUE – Any sixth power p^6 has this property, since $(p^2)^3 = (p^3)^2$; for example, $2^6 = 64 = 4^3 = 8^2$. This answer solves the Diophantine equation $x^2 = y^3$.

❹ TRUE – Any number that is not a multiple of an odd prime must be a power of 2, and if it is bigger than 2 this means it will be a multiple of 4.

❺ TRUE – Andrew Wiles proved a special case of the Taniyama-Shimura-Weil conjecture relating elliptic curves to modular forms, and this implied that Fermat's Last Theorem was true.

THE BLUFFER'S SUMMARY

Fermat's Last Theorem – that it is impossible for two integer cubes to sum to another cube and similarly for any higher power – was an open question for 350 years.

Prime numbers

'There is no apparent reason why one number is prime and another not. To the contrary, upon looking at these numbers one has the feeling of being in the presence of one of the inexplicable secrets of creation.'

DON ZAGIER

Prime numbers are the building blocks of all other whole numbers, which gives them a special fascination to mathematicians. Despite how fundamental they are, there are many aspects of the primes that we do not yet understand and many tantalizing conjectures still waiting to be proved. But you don't necessarily need to solve these deep problems in order to make the history books: hundreds of thousands of dollars are on offer simply for discovering the newest, biggest prime number.

Figuring out the secrets of primes takes real genius – finding new ones just needs patience and computing power.

1 1 is a prime number because it can only be divided by itself and 1.

TRUE / FALSE

2 Every whole number greater than 1 can be written uniquely as a product of prime numbers.

TRUE / FALSE

3 There is a biggest prime number, but we have not yet found it.

TRUE / FALSE

4 The seven biggest primes ever found are all Mersenne primes.

TRUE / FALSE

5 Pseudoprimes are composite numbers that happen to end in a 1, 3, 7 or 9 and so 'look' prime, such as 57.

TRUE / FALSE

TEN THINGS A GENIUS KNOWS

❶ What a prime number is
Prime numbers are whole numbers (integers) that are divisible by exactly two numbers: themselves and 1. For example, 7, 19 and 41 are prime, but 15 is not, because the number 15 can be divided by 3 and 5 without remainder, as well as by itself and 1. The number 1 is not prime, because it can only be divided by one number: itself. Numbers other than 1 that are not prime, such as 6 or 15, are called composite numbers, and writing a number as a product of other numbers, for example, 15 = 3 × 5, is called a factorization.

❷ Why primes are important
In a result that is so important that it is given the name 'the fundamental theorem of arithmetic', mathematicians showed that there is a unique way of writing every integer bigger than 1 as a product of prime numbers. For example, 12 = 2 × 2 × 3 and 4655 = 5 × 7 × 7 × 19. Ignoring the fact that we can write the multiplications in different orders (so 12 is also equal to 3 × 2 × 2 and 2 × 3 × 2), there are never two different collections of prime numbers that multiply together to give the same integer. This is why mathematicians think of the primes as the 'building blocks' of numbers, and is another reason why the number 1 is not considered to be prime. If 1 were prime, then the fundamental theorem would not be true because, for example, 12 would equal 2 × 2 × 3 but also 2 × 2 × 3 × 1, destroying the uniqueness of prime factorization.

❸ How many primes there are
The ancient Greek mathematician Euclid proved in around 300 BCE that there are infinitely many primes. His proof, widely considered one of the most beautiful results in mathematics, is an example of 'proof by contradiction', where we first assume that the result is false and then show that this leads to an absurd conclusion. Euclid assumed that there was a biggest prime number P. If we multiply together all the prime numbers up to P and add 1, this creates a new number N. With a few steps of logic, Euclid showed that N was either a prime bigger than P, which is impossible because P is the biggest prime, or it has a prime factorization containing a prime not in our list of primes, which is also impossible. Therefore, there cannot be a biggest prime number.

❹ The Goldbach conjecture
In 1742, the German mathematician Christian Goldbach made the conjecture that every even number (other than 2 and 4) could be written as the sum of two prime numbers. For example, 8 = 3 + 5, and 20 = 3 + 17. Nobody has yet shown this to be true, though every number of up to 18 digits has been checked and the result seems to hold. The closest result we have is a proof by Olivier Remare, which says that every even number is the sum of, at most, six primes. In 2013, the Peruvian mathematician Harald Helfgott submitted a proof of the 'weak Goldbach conjecture', showing that every odd number bigger than 5 is the sum of three primes. This is still being checked.

❺ The twin prime conjecture
Except for the primes 2 and 3, the smallest gap between primes is two. This is because all even numbers above 2 are not prime, so primes cannot be consecutive. Twin primes are those primes separated by exactly two, such as 5 and 7, 11 and 13, or 41 and 43. Primes get scarcer as we go up the number line, and yet we keep seeing these pairs of twin primes. The twin prime conjecture states that there are infinitely many of these pairs of primes. A more general conjecture, called Polignac's conjecture, predicts that there is nothing special about the number 2: there should be infinitely many pairs of primes separated by any even number.

❻ How to tell if a number is prime
It is a notoriously difficult problem to decide whether a number is prime. One method is to try dividing the number by all smaller numbers to see if there are any divisors, but this quickly gets impractical when the numbers get large. (We actually only need to check the numbers up to the square root of the number we are interested in, but even this is too much for large numbers.) There are other general methods for testing if a number is prime, but even the fastest took nine months to prove that a prime with just over 30,000 digits was prime. To find really big primes, mathematicians concentrate their searches on numbers that have special forms for which there exist specialist prime-detecting methods.

7 **What is special about Mersenne primes**
Mersenne primes are a special class of primes that can be written in the form $2^p - 1$, where p is another prime. For example, 31 is a Mersenne prime because $31 = 2^5 - 1$, and 5 is prime. There is a difficulty here, which is that not all numbers of this form are prime. So, $p = 11$ is prime, but $2^{11} - 1 = 2047$ is not prime, because $2047 = 23 \times 89$. But the special form of Mersenne numbers has enabled mathematicians to develop extremely fast methods to test whether they are prime, with the result that, in 2018, the seven biggest primes we know are Mersenne primes. The Great Internet Mersenne Prime Search (GIMPS) was founded in 1996 and uses spare processing power on personal computers to search for Mersenne primes. It found the current largest prime number in December 2017 – a Mersenne prime containing over 23 million digits, with p equal to 77,232,917.

8 **How probability is used in primality testing**
Because it is so difficult to tell whether a number is prime, mathematicians sometimes settle for 'probabilistic' methods – those that can say that a number is prime with some degree of certainty (but never 100%). For example, when we write a number in base 10 there are easy indications that it is not prime: if it ends in a 0, 2, 4, 5, 6 or 8. If it happens to end in a 1, 3, 7 or 9 then we don't know yet if it is prime, but it is more likely to be prime than a number chosen at random. By writing the number in different bases and applying simple tests in each base, we will either show at some point that it is not prime, or we will build up more evidence to suggest that it really is prime.

9 **The problem with pseudoprimes**
There are some numbers that are composite, but that look prime under the probabilistic tests described. That is, they never fail the test, no matter which number base is being looked at. These are called pseudoprimes, and there are different classes of pseudoprimes depending on the test being applied. Particular examples are Carmichael numbers, which pass a test called the Fermat primality test (based on a result first stated by Pierre de Fermat) in every base despite being composite. They can be problematic in practical applications, such as cryptography, where primes are needed but probabilistic methods must be used due to time constraints.

10 **Where primes appear in nature**
To humans, primes are of the utmost importance in helping us to understand number systems and also in helping us to develop practical applications, such as cryptography (see page 40). But nature has discovered their usefulness as well. The life cycles of cicadas are the most famous example of this. These insects stay underground for years at a time, appearing en masse to breed and begin their cycle again. Their time spent underground is almost always a prime number of years, such as 13 or 17. This is no coincidence: the primality of the life cycle makes cicadas far less likely to encounter predators when they emerge into the world. If their cycle was, say, 12 years, then predators with cycles of 2, 3, 4, 6, 8 or 10 years would coincide with the cicadas relatively often, but having a 13-year cycle makes these coincidences as rare as possible.

TALK LIKE A GENIUS

⁶ In the 1997 film *Contact*, aliens used prime numbers to communicate with humans. This is plausible, as no matter what number system aliens used, prime numbers would be a universal concept. The Arecibo message transmitted from Earth in 1974 was a pictogram of dimensions 73 × 23 – two prime numbers. ⁹

⁶ In 2013, a Chinese mathematician named Yitang Zhang showed that there were infinitely many primes separated by some number that was less than 70 million. In one of the first results of its kind, a crowd-sourced mathematics project called PolyMath managed to reduce this number to 246. Nobody knows if it can be brought down to 2 in order to solve the twin prime conjecture. ⁹

⁶ The largest known prime has over 23 million digits. The Electronic Frontier Foundation are offering prizes of $150,000 and $250,000 for the discovery of primes of over 100 million and 1 billion digits respectively. Because of community prime-searching programs like GIMPS and PrimeGrid, this prize money could easily go to the lucky armchair mathematician who has their computer running at the right time. ⁹

THE BLUFFER'S SUMMARY

Prime numbers, divisible only by themselves and 1, are the building blocks of all numbers, as every number can be uniquely written as a product of primes.

Patterns in primes

'Mathematicians have tried in vain to this day to discover some order in the sequence of prime numbers, and we have reason to believe that it is a mystery into which the mind will never penetrate.'

LEONHARD EULER

The prime numbers may form the foundation of our understanding of numbers, but they are far from being well understood themselves. Their contradictory nature has confounded mathematicians for centuries, with large-scale patterns (such as the primes becoming scarcer the further up the number line we look) competing with local unpredictability (primes very close together keep appearing even when we expect them to be far apart).

Prime distribution is a battle between order and randomness – many mathematical geniuses have tried to spot the underlying patterns.

ARE YOU A GENIUS

1 We can find a prime number that is followed by a million composite numbers before the next prime appears.
TRUE / FALSE

2 If you double the number of digits in a number, it is half as likely to be prime as before.
TRUE / FALSE

3 The sequence of numbers 7, 11, 15, 19, 23, 27, . . . contains no primes bigger than 100.
TRUE / FALSE

4 The Ulam spiral is so called because when the whole numbers are arranged in a grid, the primes appear to form a spiral.
TRUE / FALSE

5 Primes ending in 3s and 7s appear more often than primes ending in 1s and 9s.
TRUE / FALSE

TEN THINGS A GENIUS KNOWS

1 Where the next prime number will be
Being able to predict where the next prime number will be on the number line is the ultimate question about primes. The reason it is so difficult is that the size of the gaps between prime numbers is notoriously unpredictable. We know that gaps between primes can be arbitrarily long (that is, for any number we can find consecutive primes that are at least that far apart), but we also have evidence for results like the twin prime conjecture, which says that there are infinitely many consecutive primes only two numbers apart. So sometimes primes are as close together as is possible, and sometimes we have to wait a very long time before we come upon the next one.

2 How to sieve out prime numbers
One way of understanding the pattern in the primes is to keep a count of the number of primes less than a given value. So, how many primes are less than 10, or 100, or 1000? An early method used for doing this is attributed to the ancient Greek mathematician Eratosthenes and is called a 'sieve'. To find all the primes less than 100, first write out all the numbers between 2 and 100 in a list. Then circle the number '2' and cross off all multiples of 2 (since these cannot be prime). Continue the method by circling the next number not crossed off and then crossing off all multiples of that number. This will result in all the primes less than 100 being circled.

3 How to estimate the prime count
When mathematicians made tables that counted primes, they started to see a pattern emerging. It was a statistical pattern, so it was true 'on average' but it could not explain the exact count obtained in each line. This pattern was enshrined in the prime number theorem, and said that the number of primes up to a number n was roughly equal to $n/log(n)$, where $log(n)$ is the natural logarithm of the number n. Another way of saying this is that the chance of a number being prime is roughly inversely proportional to how many digits it has in it. So a number with 100 digits in it is twice as likely to be prime as a number with 200 digits in it, and ten times as likely to be prime as a number with 1000 digits in it. It shows that the primes get scarcer as we go further up the number line.

4 How accurate the prime number theorem is
For small numbers, the prime number theorem is fairly bad at giving the prime number count, but as we look at bigger and bigger numbers, the error gets smaller relative to the size of the number. For example, there are 168 numbers less than 1000 that are prime, and the theorem predicts 145. The error is 23, which is 2.3% of 1000. For a million, the error is over 6000, but this is only 0.6%, and for a billion the error drops down to 0.2%. Various improvements to the theorem can reduce the overall error, but it is the Riemann hypothesis that attempts to explain how the error behaves on small scales. (See page 36).

5 Whether there are primes in sequences
In the sequence of numbers 4, 7, 10, 13, 16, 19..., we see some primes appearing. Will there be an infinite number of primes in this sequence, or will they run out? What about if we had made a different sequence by starting with a different number, or by adding a different number each time? The mathematician Johann Peter Gustav Lejeune Dirichlet answered this question by showing that, if the first term in the sequence and the number added each time to generate the sequence are coprime (that is, there is no number other than 1 that evenly divides both of them), then the sequence will contain infinitely many primes. For example, in the sequence above the first term is 4 and the number added each time is 3, so this sequence will have infinitely many primes in it because 4 and 3 are coprime.

6 If there is a formula for prime numbers
Dirichlet's result on primes in sequences raised the question of whether there might be a number sequence that contained only primes, and therefore whether there was a formula for generating the primes. The number sequence above is generated by the formula $4 + 3n$ as we let n run through the whole numbers, but though it produces infinitely many primes, it also produces infinitely many composite numbers. Sadly, if there is such a formula, it will not be as simple as a number sequence. People have shown that there will never be a polynomial formula in one variable for outputting primes, though a complicated polynomial with ten variables has been found whose positive values are exactly the prime numbers.

7 The patterns in the Ulam spiral

The Polish mathematician Stanislaw Ulam was doodling one day in a boring seminar, and discovered a striking pattern in the prime numbers. When the integers were arranged in a spiral, he noticed that the primes tended to line themselves up along diagonals. His arrangement was called the Ulam spiral, and showed that the primes were not distributed randomly among the numbers. In particular, it showed that some quadratic formulae contained far more primes in them than others did.

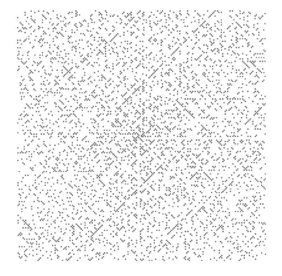

8 The relationship between primes and quadratics

A quadratic formula is one of the form $ax^2 + bx + c$ where a, b and c are specific numbers. We learn about them in school because they are used to predict the motion of projectiles (among other things), but they seem to have an uncanny relationship with the prime numbers, too. It is still unknown whether any quadratic formula can output infinitely many primes, the way Dirichlet's number sequences do, but the British mathematicians Hardy and Littlewood conjectured a result that explains why some values of a, b and c generate far more primes than other choices do. For example, the quadratic $x^2 + x + 41$ is prime for all the whole number values of x from 0 to 39. This explains one of the diagonal lines in the Ulam spiral and relates to some incredibly deep mathematics about number systems.

9 How random the primes appear to be

A prime number (other than 2) cannot end in a 0, 2, 4, 6 or 8 because even numbers cannot be prime, and a prime number (other than 5) cannot end in a 5 because then it would be divisible by 5. So primes always end in a 1, 3, 7 or 9. An obvious question to ask is whether primes take each of these four endings equally often, or whether they are more likely to end in some numbers than others. If we make running totals of the number of primes ending in each of the four digits, 3 and 7 appear to be 'ahead' almost all of the time. Yet 1 and 9 will each catch up in the race infinitely often. Attempts to explain this phenomenon make use of the generalized Riemann hypothesis, which has not yet been proven.

10 How primes avoid each other

Research published in 2016 showed further evidence that the primes do not appear to behave entirely randomly. Mathematicians Kannan Soundarajan and Robert Lemke Oliver found evidence to suggest that consecutive primes are less likely to have the same last digit than would be expected. A prime ending in a 1 had a 30% probability of being followed by a prime ending in a 3 or 7, 22% by a 9 and only 18% by another 1. The same thing happened with the other digits, so that a prime ending in a 3 was least likely to be followed by another prime ending in a 3. This happens in other number bases, too. For example, a prime number when divided by 6 either has remainder 1 or remainder 5. If a prime has remainder 1, then the next prime is far more likely to have remainder 5: primes seem to avoid being near other similar primes. The explanation for this is still an open problem.

TALK LIKE A GENIUS

⁶ The prime number 73,939,133 is especially prime. If you keep removing the right-most digit of this number, the remaining numbers are all prime, too. It is currently the largest known number with this property. ⁹

⁶ The number 24 has an intimate connection with the squares of prime numbers. If you square any prime number bigger than 3 and take away 1, the answer will be a multiple of 24. Or if you square any two prime numbers bigger than 3, the difference of these numbers will be a multiple of 24. ⁹

⁶ The prime number theorem was first conjectured, in 1792, by Carl Friedrich Gauss at the tender age of 15. He found the result by examining tables of prime number counts, but did not have a proof of his conjecture. It took over 100 years before proofs were independently found by Jacques Hadamard and Charles-Jean de la Vallée-Poussin, in 1896. ⁹

WERE YOU A GENIUS?

1 TRUE – We can find consecutive prime numbers separated by as big a number as we can think of.

2 TRUE – The probability of a number being prime is inversely proportional to the number of digits in it.

3 FALSE – Any arithmetic sequence where the number being added each time (in this case, 4) is coprime to the first number (in this case, 7) will contain infinitely many primes.

4 FALSE – In the Ulam spiral the numbers are arranged in a spiral, and then the prime numbers appear to be arranged along particular diagonals.

5 FALSE – In the long run, primes end in each of the digits equally often, although along the way the 3s and 7s appear to be ahead in the race more of the time.

THE BLUFFER'S SUMMARY

Primes get scarcer up the number line in a statistically predictable way, though their distribution on small scales is still a mystery, meaning that we cannot predict where the next prime will be.

The Riemann hypothesis

'If you could be the Devil and offer a mathematician to sell his soul for the proof of one theorem [...] I think it would be the Riemann hypothesis.'

HUGH MONTGOMERY

Probably the most famous unsolved problem in mathematics, the Riemann hypothesis is one of the seven Millennium Prize Problems worth one million dollars to whoever can crack its puzzle. It brings together two apparently unrelated fields – complex analysis and number theory – theorizing that the placement of the prime numbers is governed by the behaviour of imaginary numbers.

Whoever proves the Riemann hypothesis is unlikely to be motivated by the million-dollar prize, but by a desire to understand the beauty of the solution and its wide-ranging implications on mathematics.

ARE YOU A GENIUS
?

1 Adding up $1 + \frac{1}{4} + \frac{1}{9} + \frac{1}{16} + \ldots$, where the denominators are all the square numbers, gives an answer of infinity.

TRUE / FALSE

2 A (true) consequence of Riemann's work is that $1 + 2 + 3 + 4 + \ldots = -\frac{1}{12}$.

TRUE / FALSE

3 Any infinite sum of numbers can be written as an infinite product involving only prime numbers.

TRUE / FALSE

4 If the Riemann hypothesis is true, it will explain why prime numbers are not found in the places that mathematicians predict them to be.

TRUE / FALSE

5 The Riemann hypothesis being proven will break all modern cryptography methods.

TRUE / FALSE

TEN THINGS A GENIUS KNOWS

1 **What the Riemann zeta function is**
The Riemann zeta function, denoted by $\zeta(s)$, takes a number s and finds the sum of all the fractions whose numerator is 1 and whose denominator is a positive whole number raised to the power of s. In mathematical notation this is written

$$\zeta(s) = \frac{1}{1^s} + \frac{1}{2^s} + \frac{1}{3^s} + \frac{1}{4^s} + \cdots$$

The formula is a sum of an infinite number of things but, using the ideas of analysis (see page 144), can often be shown to give a sensible answer. When $s = 1$ we get something called the harmonic series, which is infinite, but when $s = 2$ the answer, quite amazingly, turns out to be $\pi^2/6$.

2 **The relationship of the zeta function to prime numbers**
At first glance, the zeta function doesn't appear to have any connection with the primes at all, as it is a sum involving all of the natural numbers. But mathematician Leonhard Euler showed that the formula could be rewritten in terms of the primes. Instead of writing the zeta function as a *sum* of fractions, he showed that it could be written as a *product* of fractions where the denominators are $(1 - p^{-s})$ as p runs through each of the primes in turn. This equivalence is a result of the fundamental theorem of arithmetic, which tells us that every number can be written uniquely as a product of prime numbers. It means that anything we learn about the zeta function has something to tell us about the primes. For example, knowing that $\zeta(1)$ is infinite tells us that there must be infinitely many primes.

3 **How to define the zeta function on complex numbers**
The zeta function only makes sense for numbers s that are bigger than 1. For example, if we put $s = 0$, then the zeta function would give us $1 + 1 + 1 + 1 + \ldots$ which is infinite. Riemann's genius idea was finding a sensible way to define the zeta function for numbers less than 1 – that is, to give finite answers – and also showing how the function could give sensible answers when s is a complex number. The way he did this was by finding another function that gave exactly the same answers as $\zeta(s)$ when $s > 1$, and using this function to tell us the answers to the other numbers where $\zeta(s)$

was annoyingly infinite. This method is called 'analytic continuation' and it suddenly opened up a whole new way to think about the zeta function.

4 **The strange results that the zeta function gives us**
The analytic continuation of the Riemann zeta function is uniquely defined (that is, if Riemann had tried to extend the zeta function by using some other function instead, it would give the same answers) and makes sense mathematically, but it also delivers some head-scratching results. For example, it says that $\zeta(0)$, which we've seen should be $1 + 1 + 1 + 1 + \ldots$, is equal to $-1/2$. It says that $\zeta(-1)$, which is $1 + 2 + 3 + 4 + \ldots$, is equal to $-1/12$. These seem to be ridiculous results, but they are actually useful in physics, in areas such as string theory and quantum physics. The zeta function continuation also tells us that there are some numbers that give zero when inputted into the function. Every negative even number has this property, and these are called the 'trivial' zeros, but there are infinitely many other zeros that are more complicated than this.

5 **What the Riemann hypothesis says**
The Riemann zeta function takes a complex number, which we can think of as x- and y- coordinates, and outputs a number. Which coordinates give an output of zero, apart from the trivial zeros we have already seen? Mathematicians have shown that all the numbers that give zero have their x-coordinate between 0 and 1. The Riemann hypothesis states that all the coordinates that give zero have their x-coordinate exactly equal to $1/2$. Plotting these coordinates on the complex plane (that is, an x–y coordinate grid), this is the same as saying that the zeros all lie along a single line, known as the critical line. Mathematicians have checked over a trillion zeros and they all lie on this line, but it is still possible that the next zero to be found will show the conjecture to be wrong.

6 **Why the Riemann hypothesis is important**
We have seen (page 33) that there is a formula governing the statistical distribution of the prime numbers. This result, called the prime number theorem, says that the average gap between primes up to a number n is roughly $log(n)$. This gives us a rough idea of where the primes are likely to be and how many of them there are, up to a given number, but it does not tell us *exactly* where each prime is.

Riemann was able to show that the zeros of the zeta function govern the error between where a prime is predicted to be by the prime number theorem, and where it actually is. The zeros are somehow controlling the small-scale fluctuations of the primes – an incredible result that links together the building blocks of numbers with the strange world of imaginary numbers.

7 Whether the Riemann zeta function is special

Instead of trying to prove the Riemann hypothesis directly, some mathematicians wondered whether there was a larger class of functions that exhibited similar phenomena to the Riemann zeta function. By investigating this wider collection of functions, they hoped to pin down exactly which properties of the zeta function were really important in making the conjecture work. What they found were 'Dirichlet L-functions', which were infinite sums with the same denominators as the Riemann zeta function but with different numerators. The numerators of an L-function obey a form of the fundamental theorem of arithmetic. So, the numerator above 15^s would be equal to the product of the numerators above 3^s and 5^s. These L-functions also appear to have their zeros lying on the critical line of $x = 1/2$, and the conjecture that this is always so is called the generalized Riemann hypothesis.

8 Whether the Riemann hypothesis will help crack cryptography

If the Riemann hypothesis were true, this would have profound consequences for our understanding of prime numbers, but it would also answer open questions in other areas of mathematics. One practical consequence is about primality testing. There are currently very few algorithms that can test whether numbers are prime in a short amount of time (that is, polynomial time). One potential method is called Miller's test, but the proof that it is guaranteed to work relies on the generalized Riemann hypothesis. However, the understanding of primes and primality testing that the Riemann hypothesis would give us would have no impact on current cryptographic methods, because it tells us nothing about how to factor numbers that are products of primes.

9 How to use the Riemann hypothesis without knowing if it is true

A growing number of results are being proven using the ingenious method of first assuming the Riemann hypothesis to be false, and then assuming it to be true, and showing that the result in question is true either way. One concerns the error in the prediction of the number of primes below a given value. For the first 10^{23} numbers, the prime number theorem consistently overestimates how many primes there are, but the mathematician Littlewood showed, using this method of first assuming the Riemann hypothesis true and then false, that the prime number theorem would switch between over- and underestimations an infinite number of times.

10 The current status of the Riemann hypothesis

British mathematicians Hardy and Littlewood proved that there are infinitely many zeros of the zeta function on the critical line $x = 1/2$. Another mathematician named Conrey has proven that at least two-fifths of the zeros lie on the line. Every zero that has ever been found lies on the critical line. Similar results have been proved about other functions or in other number fields, and predictions made by the Riemann hypothesis have all proven to be true. At the same time, we have examples where counterexamples to such conjectures have been found at extremely high numbers – so high that no computer in existence could find such a number. The current situation is therefore that most mathematicians believe the Riemann hypothesis to be true, but a few still believe it may turn out to be false.

TALK LIKE A GENIUS

❝ Alan Turing was fascinated by the Riemann hypothesis and started designing a machine with a complicated series of gears that could compute the zeros of the zeta function. His work was interrupted by the Second World War, where he used the skills he had developed to create a computer to crack the German Enigma code. Returning to the problem in Manchester in 1950, he became the first person to calculate Riemann zeros using an electronic computer. ❞

❝ Riemann died from tuberculosis at only 39 years of age. On hearing of his death, his housekeeper went through his office throwing away much of the clutter that was there. But Riemann was both messy and a perfectionist, meaning that he had many pages of scribbles that he hadn't yet deemed good enough to publish. Before any other mathematicians could get to the house, the housekeeper had resigned hundreds of pages to the fire. Given that the whole Riemann hypothesis stemmed from a single nine-page paper, we can only imagine what treasures were lost that day. ❞

❝ In 2017, a spoof article was published under the name of Donald Trump, proving the Riemann hypothesis by using 'alternative facts'. The same article also proved that the circle has 11 sides and that 0 = 1. ❞

WERE YOU A GENIUS?

1 FALSE – The sum is finite and, amazingly, has a value of $\pi^2/6$. (This answer is the Riemann zeta function $\zeta(s)$ evaluated at $s = 2$.)

2 TRUE – Amazingly, this is true and has been used to make predictions in physics.

3 FALSE – Only certain infinite sums have this property.

4 TRUE – The zeros of the Riemann zeta function govern the error between where the primes are predicted to be by the prime number theorem and where they really are.

5 FALSE – While the Riemann zeta function tells us about the placement of primes, it does not provide any practical methods for factoring numbers, which is what modern cryptography depends on.

THE BLUFFER'S SUMMARY

The zeros of the Riemann zeta function are conjectured all to lie on the same line, and their placement governs where the prime numbers will be found.

Keeping secrets with primes

'No discovery of mine has made, or is likely to make, directly or indirectly, for good or ill, the least difference to the amenity of the world.'

G.H. HARDY

Mathematician G.H. Hardy was proud that his work in number theory would never be used to influence the world in any way, but would remain pure abstract logic. Fast forward to the 21st century and modern society would not exist without his, and others', work on primes and a special type of clock arithmetic. It underpins every piece of internet communication, every purchase on a credit card, every song bought online, and every piece of data stored securely on a computer.

Prime numbers and clock arithmetic keep our data safe, but could some genius find a way to outwit these mathematical algorithms?

1 In public key cryptography people can send encrypted messages without first needing to agree on a mutual decryption key.

TRUE / FALSE

2 A one-way problem is a calculation that is only possible in one direction, and impossible to reverse.

TRUE / FALSE

3 It is extremely difficult to factor any large number.

TRUE / FALSE

4 The numbers 3 and 10 are considered the same when working modulo 7.

TRUE / FALSE

5 The RSA encryption method has already been cracked and is unsafe to use.

TRUE / FALSE

TEN THINGS A GENIUS KNOWS

1 **The problem with regular cryptography**
Cryptography is the art of altering messages and hiding their content so that third parties cannot read them. One easy example is the Caesar cipher, in which the alphabet is shifted by a certain amount so that each letter is replaced with a new letter – for example, A becomes C, B becomes D, C becomes E and so on, turning MATHS into OCVJU. In a more physical example, the message may be put in a locked box, to which only the sender and recipient have keys. The problem with these kinds of encryption is that the sender and receiver have to swap keys to the message in advance (either a physical key or a clue to decrypting the message), and at this point a wily eavesdropper could get a copy of the key and have access to all subsequent messages.

2 **What public key cryptography is**
Public key cryptography provides ways of exchanging messages securely without needing to swap keys in advance. The mailbox of your house is one example: anybody can put a message in your box, but the messages are secure because only you have the key to the mailbox. A more interesting example is called the 'Three Pass Protocol'. Alice wants to send a secret message to Bob. She puts her message in a box and locks it with a padlock before sending it to Bob. Bob attaches his own padlock and sends the box back to Alice. She removes her padlock and sends the box to Bob one last time, who removes his own padlock and reads the message. The box is always locked as it is being passed, so no third party has access, but Alice and Bob do not ever need to swap keys to their padlocks.

3 **How to do cryptography with maths**
The equivalent of a padlock in mathematics is called a one-way problem. This is a problem that is very easy to do one way (like closing the padlock) but very hard to reverse without a key. Modern cryptography is based on two particular one-way problems. The first of these is factoring numbers. For example, it is easy to multiply two numbers together and get 87,713, but it is hard to go backwards from the answer to find the two numbers. The important thing here is that the two numbers are large primes. A number like 1040 would be easy to factorize

because it is very obviously divisible by 2 and 5, but it's not obvious at all where to begin with 87,713. The only technique here is to try dividing primes until you find the correct one, and when the numbers are big, this takes a very long time. But it is easy with a key: if you know that one of the factors is 239, you can easily find the other one.

4 **What modular arithmetic is**
Describing the second one-way problem used in cryptography requires an understanding of an area of mathematics called modular arithmetic. Modular arithmetic assumes that numbers are arranged in a circle as on a clock face, and that every time we go around the circle we start again at 0. When 12 numbers are on the clock face this means that 13 o'clock is the same as 1 o'clock and 20:00 is the same as 8:00. If seven numbers are on the clock face, then 9 o'clock would be the same as 2 o'clock. If the clock has n numbers, we say that we are working 'modulo n'.

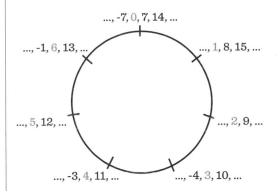

..., -7, 0, 7, 14, ...

..., -1, 6, 13, ...

..., 1, 8, 15, ...

..., 5, 12, ...

..., 2, 9, ...

..., -3, 4, 11, ...

..., -4, 3, 10, ...

5 **What the discrete logarithm problem is**
The second one-way problem used in cryptography is called the discrete logarithm problem, and is based on the difficulty of taking logarithms when working with modular arithmetic. A logarithm is the reverse of exponentiation, which means taking powers. An example of a logarithm is to ask: which power of 2 gives 32? In other words, what is x in the equation $2^x = 32$? Standard logarithms can be found easily, but the problem becomes surprisingly difficult once we start using modular arithmetic. A typical discrete logarithm problem might ask for the power

of 2 that gives an answer of 15 (modulo 19). Just as with factoring, there is no easy way to get the answer, except to try raising 2 to different powers in turn until the correct answer is found. And the problem is at its most difficult when the modular arithmetic is done with a prime number.

6 The RSA algorithm
The most widely used algorithm for sending encrypted messages is called RSA, named after its creators Ron Rivest, Adi Shamir and Leonard Adleman. The method starts with Alice making two numbers, n and e, public, which Bob can use to encrypt messages. Alice's private key consists of two prime numbers p and q which, together, multiply to make n, and another number d which, when multiplied with e, gives an answer of 1 (modulo $(p-1)(q-1)$). The difficulty in factoring means that knowing n doesn't allow anyone to figure out the private numbers p and q, and d is also safe unless the eavesdropper knows p and q. Bob encrypts a message m by computing m^e (modulo n) (which is safe because of the difficulty of the discrete logarithm problem) and Alice decrypts it by raising it to the power d.

7 Where RSA is used
You have probably already used the RSA algorithm many times today without realizing it. If you have been on a website such as Facebook, whose address starts with https://, then you have used RSA, because you are using a secure protocol to check the authenticity of the website. If you have used a fingerprint or PIN to access your Android phone, then you have used RSA. If you have bought an item online with a credit card, then you have used RSA to encrypt your details. The RSA encryption method is not the only way to encrypt data, but it is still the most widely used for data sent over the internet.

8 Things to be aware of when using RSA
The best choices for the primes p and q are numbers that are fairly similar in size, with one having just a few more digits than the other. This is the most difficult combination to factor. The primes should each be at least 100 digits long and ideally more than 300 digits to keep up with modern advances in computing. The number e should be large enough so that the logarithm problem cannot be solved by standard (that is, non-modular-arithmetic-based) methods, but small enough so that the encryption process is efficient.

9 The importance of randomness
An attack on RSA by researchers in 2012 managed to crack 0.2% of all codes by noticing a very simple thing about how people were choosing their prime numbers. If two people using RSA happen to pick one of the same primes to compute their public key n, then their choice of primes can easily be found by comparing their public keys. Such a thing should not happen if the prime numbers are chosen completely at random, but the random number generators being used had subtle flaws, which meant that the numbers were not as random as they should have been. Another use of randomness is the idea of padding: introducing random noise into the message to stop people from trying to crack the code by simply guessing at what the message contains.

10 Whether RSA has been cracked already
It is unknown whether either the discrete logarithm problem or the factoring problem are solvable in polynomial time (that is, fast) on standard computers. The most recent attack on the discrete logarithm problem took place in 2015 with a program called Logjam, developed by a group of computer scientists. By pre-computing data for a number of 512-bit primes that were commonly used by RSA, they were able to crack the codes in a matter of minutes. Doing these computations for 1024-bit primes would cost millions of dollars, but it would be feasible for a large organization, such as the National Security Agency (NSA), which is why 2048-bit primes are currently recommended.

TALK LIKE A GENIUS

❝ The RSA algorithm was released in 1977, but it turned out that a British mathematician named Clifford Cocks had developed the same method already in 1973. Since Cocks was working for the secret government communications agency GCHQ at the time, his work was not declassified and made public until 1997. ❞

❝ The RSA numbers are a list of 54 numbers that are the product of two similar large primes. The list was created in 1991 to encourage the mathematical community to develop methods to factorize them, with prizes of up to $200,000 for the solutions. The largest RSA number to be factored – a 768-bit number – took two years of computation on hundreds of computers and a total of 1500 CPU years. Though the challenge ended in 2007, only 19 of the 54 problems were solved and so mathematicians are still attempting to factorize the rest of the list. ❞

❝ It has been shown that a quantum computer would be able to crack the RSA code efficiently. Although sufficiently powerful quantum computers aren't expected to make an appearance any time soon, researchers are already working on creating "post-quantum cryptography" to replace RSA should this ever happen. ❞

WERE YOU A GENIUS?

1 TRUE – A 'public key', which is available to everyone, is used to encrypt messages, while a 'private key', which is kept secret, is used to decrypt messages.

2 FALSE – A one-way problem is easy to do in one direction but difficult to reverse without a 'key' to the problem.

3 FALSE – Some large numbers are easy to factor but those that are products of two similarly sized primes are incredibly difficult.

4 TRUE – 3 and 10 are separated by 7, meaning they are at the same position on a clock face with seven numbers.

5 FALSE – RSA is probably compromised for small primes (less than 1024 bits), but there is still no way to crack it for larger ones.

THE BLUFFER'S SUMMARY

The mathematics of prime numbers ensures that our data is kept safe online, though the codes could be cracked if a genius could figure out how to factorize numbers quickly (or build a quantum computer).

The Birch and Swinnerton-Dyer conjecture

'Mankind is not a circle with a single centre but an ellipse with two focal points of which facts are one and ideas the other.'

VICTOR HUGO

How far does the Earth travel around the Sun in a week? A simple question about the lengths of arcs in ellipses turned out to be fiendishly difficult to solve, and gave rise to the study of geometric shapes called elliptic curves. These have remarkable algebraic properties, which are exploited today by cryptographers keeping our data safe. The Birch and Swinnerton-Dyer conjecture predicts a deep connection between elliptic curves, prime numbers and imaginary numbers, and is considered so important that it is one of Clay Mathematics Institute Millennium Prize Problems, worth one million dollars.

Can any genius out there prove that Birch and Swinnerton-Dyer's predictions are correct?

1 There is no simple formula for the circumference of an ellipse.

TRUE / FALSE

2 A circle of radius 1 centred at (0,0), having equation $x^2 + y^2 = 1$, does not go through any coordinates (x,y) where x and y are both rational numbers (except where x or y are equal to 1).

TRUE / FALSE

3 Elliptic curves were known to the third-century mathematician Diophantus of Alexandria.

TRUE / FALSE

4 The Birch and Swinnerton-Dyer conjecture relies on the proof of Fermat's Last Theorem.

TRUE / FALSE

5 Elliptic curve cryptography is more efficient than Rivest-Shamir-Adleman (RSA) cryptography, one of the main encryption standards used.

TRUE / FALSE

TEN THINGS A GENIUS KNOWS

① How elliptic curves got their name

At the heart of the Birch and Swinnerton-Dyer conjecture are objects called elliptic curves, which are equations of the form $y^2 = x^3 + ax + b$ where a and b are numbers chosen so that $x^3 + ax + b$ has three distinct roots. Another way of phrasing this is that an elliptic curve is the square root of a cubic polynomial. Elliptic curves are not equations of ellipses, so where does the name come from? Ever since Kepler had shown that the planets move in ellipses around the Sun, mathematicians and scientists had been trying to find formulae for things like the area and arc-length of an ellipse, but they turned out to be far more difficult to work with than circles. Finding the arc-length involved integrating the square root of a cubic polynomial, and not even the most genius mathematicians could figure out how to do this.

② How elliptic curves are related to Diophantine equations

To be able to integrate the square root of a cubic polynomial, mathematicians needed to be able to write this equation in terms of objects called rational functions. But at some point they realized that doing so would violate other results, like the fact that the equation $x^4 + y^4 = z^4$ had no positive integer solutions (a consequence of Fermat's Last Theorem). These investigations created a link between elliptic curves and Diophantine equations (polynomial equations in which only integer solutions are sought). Such equations were described by Diophantus of Alexandria in the third century, who had also studied elliptic curves and may have had an inkling of their unusual algebraic properties.

③ The group law of elliptic curves

Elliptic curves can be thought of as Diophantine equations, meaning that we can ask whether they have any integer solutions. (Mathematicians normally talk about finding rational points – solutions that are integer fractions – since having a rational solution translates into finding an integer solution and vice versa.) Diophantus himself noticed that if two rational solutions were already found, then these could be combined together in a clever way to find a third rational solution. This observation brought to light the most important property of elliptic curves: that their

points form an algebraic object called a group (see page 116), in which any two points can be 'added' to obtain a third point. This addition is called the group law of elliptic curves.

④ How to explain the group law geometrically

The 'addition' of points on elliptic curves has a geometrical interpretation. In order to add together points P and Q on an elliptic curve, draw a straight line through P and Q and see where this line hits the curve again. If the third point is called R, then the group law is that P + Q = –R, that is, the reflection of R in the horizontal axis. If P and Q are directly above one another then P + Q = 0, with this 0 thought of as the 'point at infinity'. The group law only works when the cubic equation $x^3 + ax + b$ has three distinct roots, which is why only these cubics are considered to form elliptic curves.

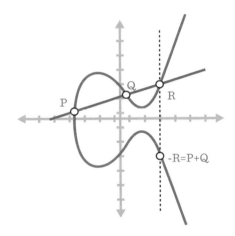

⑤ What the rank of an elliptic curve is

If we already know some rational points on an elliptic curve, we can use the group law to find more of them. In 1922, Louis Mordell proved a striking theorem: every elliptic curve has a finite collection of rational points from which all other rational points can be generated. These special points are called a basis for the rational points on the curve. We can take a point in the basis and keep adding it to itself over and over again. If we eventually get back to the point at which we started, we say this point has finite order. But if we keep generating new points forever, we say it has infinite order. The rank of the elliptic curve is the number of points in the basis that have infinite order. If the rank is 0, then the elliptic curve only has finitely many rational points.

6 How to estimate the rank using prime numbers

Mordell's theorem told mathematicians that the rank of an elliptic curve was always a finite number, but it did not tell them how to calculate this number. In the 1960s, Bryan Birch and Peter Swinnerton-Dyer were running computer programs to calculate the answer to a simpler problem: finding the number of rational points N_p on an elliptic curve modulo a prime number p. (See page 41 to learn about modular arithmetic.) When working modulo a prime number, there are only finitely many points to check, so this is computationally straightforward. By multiplying together N_p/p for more and more primes, they were able to do a calculation that gave better and better estimates for the rank of the elliptic curve, and they conjectured that this was not a fluke.

7 Defining an L-function from an elliptic curve

When there is a product involving a formula defined only for prime numbers, there is a good chance that this will be related to an L-function. This means that the product over primes can be transformed into an infinite sum involving natural numbers. (The Riemann zeta function in the Riemann hypothesis is another example of an L-function – see page 37.) The L-function for an elliptic curve E is usually denoted by $L(E,s)$, where s is a complex number. Its formula should only make sense for particular complex numbers, but work on the proof of Fermat's Last Theorem by Andrew Wiles and others showed that the formula actually made sense for all complex values of s.

8 What the Birch and Swinnerton-Dyer conjecture predicts

Birch and Swinnerton-Dyer's early conjecture about how to find the rank of an elliptic curve using prime numbers led them to a far more sophisticated conjecture about the behaviour of $L(E,s)$. The conjecture worth one million dollars states that $L(E,s)$ has a zero of order r at $s = 1$, where r is the rank of the elliptic curve E. Another way of stating

this is that $L(E,s)$ can be written as an infinite sum of powers of $(s - 1)$ with the smallest power being the rank of the elliptic curve. A consequence of this conjecture is that, if we compute $L(E,1)$ and get an answer of 0, then the elliptic curve E has infinitely many rational points.

9 Whether the Birch and Swinnerton-Dyer conjecture is likely to be true

The original Birch and Swinnerton-Dyer conjecture was based mostly on numerical evidence, rather than mathematical intuition. We now know that the conjecture is definitely true for elliptic curves of rank 0 and 1, but nothing is known for curves of higher rank. Even checking that the conjecture is true for special examples of elliptic curves is difficult, because current methods can only find the exact order of the zero at $s = 1$ when we already assume that the order is less than 4. A proof of the full conjecture therefore seems a long way off.

10 Why we care about elliptic curves

The Birch and Swinnerton-Dyer conjecture is providing information about the structure of the group at the heart of every elliptic curve. This algebraic structure is important to all of us because it is used in modern cryptography to encrypt sensitive data, such as passwords, credit-card information and digital signatures. It has an advantage over the standard method of encryption, RSA, because it requires a smaller key size to achieve the same level of security. To use elliptic curve methods we must be able to count the points on elliptic curves modulo primes, and since this is difficult, agencies publish details of particular elliptic curves that are recommended for cryptography.

TALK LIKE A GENIUS

◦ In the court of Holy Roman Emperor Frederick II, Fibonacci was challenged with the problem of deciding whether the number 5 could be the area of a right-angled triangle. It can: a right-angled triangle with side lengths of $^{20}/_3$, $^3/_2$ and $^{41}/_6$ does the trick. But, for a general integer, the problem remains unsolved; that is, unless the Birch and Swinnerton-Dyer conjecture is true. ◦

◦ In 2013, revelations emerged through the whistle-blower Edward Snowden that a particular elliptic curve algorithm used as a national standard for encryption had a backdoor that would allow those who knew about it to crack the code. The National Security Agency (NSA) not only knew about this backdoor but had pushed for the algorithm to be included. Since then some have questioned the security of elliptic-curve-based encryption methods. ◦

1 TRUE – The integral involved in the calculation of the circumference (or arc-length) of an ellipse has no answer in terms of elementary functions.

2 FALSE – For example, take $x = ^3/_5$ and $y = ^4/_5$.

3 TRUE – Diophantus knew of elliptic curves, and even of the algebraic structure that their rational points possessed.

4 TRUE – The proof of Fermat's Last Theorem showed that the L-function used in the Birch and Swinnerton-Dyer conjecture made sense for all complex numbers.

5 TRUE – Key lengths are generally shorter than the corresponding RSA keys providing the same level of security.

THE BLUFFER'S SUMMARY

The Birch and Swinnerton-Dyer conjecture relates rational points on elliptic curves to infinite expressions of prime numbers and complex numbers.

Proofs

'You don't have to believe in
God, but you should believe
in The Book.'

PAUL ERDŐS

The Hungarian atheist mathematician Paul Erdős
believed in a book in which God had written the
most elegant versions of every proof in mathematics.
To discover a proof 'from the book' is the source
of the greatest joy for any mathematician. A
proof should not only be a way of confirming that
something is true, but it should provide insights into
why the result is true and find new connections in
the web of mathematical knowledge. Many different
methods of proof have been discovered, with some
causing fierce debate and a questioning of the very
basics of logical thought.

**Proof remains at the very heart of mathematics
– nothing is believed until it has a proof.**

1 A proof of a mathematical result
is any plausible argument that
convinces people of its correctness.
TRUE / FALSE

2 The negation of 'all sheep have
four legs' is 'no sheep have four
legs'.
TRUE / FALSE

3 The law of the excluded middle
says that either a statement or its
negation must be true.
TRUE / FALSE

4 The contrapositive of 'if I am
drunk then I have had alcohol' is
'if I am not drunk then I have not had
alcohol.'
TRUE / FALSE

5 Proof by induction is a
method used when showing that
a statement is true for each natural
number.
TRUE / FALSE

TEN THINGS A GENIUS KNOWS

1. The language of logic

A proof is a sequence of logical deductions, starting from statements that are known to be true and ending by deducing that new statements are true. For example: Genghis is a cat; every cat has nine lives; therefore Genghis has nine lives. Statements like 'Genghis is a cat' are called propositions, and assert something about the subject of the sentence. The negation of a proposition is the statement that denies it. So the negation of 'every cat has nine lives' is the statement 'not every cat has nine lives'. This is equivalent to 'there is a cat who does not have nine lives' (*not* 'no cat has nine lives'). The phrases 'for all/every' and 'there exist' are called quantifiers.

2. The symbols of logic

The study of logic uses a collection of symbols that simplify statements and highlight the logical structure of deductions. A proposition is usually denoted by a letter, such as P; the quantifiers 'for all' and 'there exist' are written as $\forall$ and $\exists$; negations are written using $\neg$, so 'not P' is written '$\neg P$'; the logical connectives 'and' and 'or' are written using the shorthand $\wedge$ and $\vee$; and 'therefore/implies' is written as $\rightarrow$. A statement like $P \wedge \neg Q \rightarrow R$ would be read as 'P and not Q implies R'; for example, (it is raining) and (I do not have an umbrella) implies that (I am wet).

3. The laws of thought

The earliest formal theory of logic was developed by the ancient Greek philosopher Aristotle, and his principles were later encapsulated as three laws by Bertrand Russell in 1912. First, the law of identity says 'Whatever is, is' – in symbols, $\forall P : P \rightarrow P$. Second, the law of non-contradiction says 'Nothing can both be and not be' – in symbols, $\neg(A \wedge \neg A)$. Thirdly, the law of the excluded middle says 'Everything must either be or not be' – in symbols, $A \vee \neg A$. Another way of saying the third law is that either a statement or its negation must be true. In systems of logic where everything is either true or false, the second and third laws are equivalent to each other.

4. How to prove a theorem by contradiction

Proof by contradiction, also called *reductio ad absurdum*, makes use of the law of the excluded middle, using the result that either a statement or its negation is true. So, to prove a statement by contradiction, a mathematician will assume that its negation is true and try to reach an absurdity in the argument, proving that the negation is, in fact, false. For example, to prove the statement 'There is no biggest number', we first assume its negation: 'There is a biggest number'. Call the biggest number N. But then $N + 1$ is a bigger number, which contradicts our assertion that we already found the biggest number. Therefore, the negation is false, and so the original statement must be true.

5. The method of infinite descent

The method of infinite descent is a particular type of proof by contradiction that uses the fact that there is a smallest positive integer. The idea is to show that, if an example of something existed, then this would imply the existence of a smaller solution, which would, in turn, give a smaller solution, continuing so as to show an infinity of smaller and smaller solutions. If each solution is indexed by a natural number, then this conclusion is a contradiction. The Greeks used a geometric proof by descent to show that $\sqrt{2}$ was irrational, and Fermat used the technique often to prove results about Diophantine equations.

6. What constructivism is

Proof by contradiction shows that something is true by proving that it is not false. It is an example of a non-constructive proof method, and is one by which some mathematicians are unconvinced. Constructivism is a branch of mathematical philosophy that rejects the law of the excluded middle and holds that statements are only true if they can be proved directly. For example, suppose we wish to prove that, in the decimal expansion of π, there is a digit that occurs infinitely often. Its negation is easy to refute: if each digit occurred finitely many times then π would be a rational number, which it is not. But to prove the result directly, a constructionist would need to exhibit a digit that occurred infinitely often, and nobody yet knows how to do this.

7 What the contrapositive is

A more direct method of proof than contradiction is the 'contrapositive'. This uses the fact that $P{\rightarrow}Q$ is logically equivalent to $\neg Q{\rightarrow}\neg P$. That is, if P implies Q, then it is also true that not-Q implies not-P. For example, asserting 'If it is Tuesday, then I have to go to work' is logically the same as saying 'If I do not have to go to work, then it cannot be a Tuesday'. This equivalence can be seen in the diagram with two circles, where the dashed circle represents all situations in which P is true, and the solid circle represents all situations in which Q is true. Saying that 'P implies Q' is saying that the dashed circle sits inside the solid one, which is the same as saying that anything not in the solid circle is also not in the dashed circle.

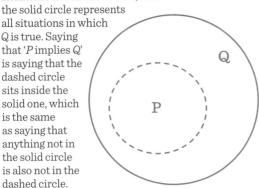

8 How to prove something by induction

Induction is a common method of proof when proving statements about natural numbers. It uses two facts: that there is a 'first' natural number (that is, 1) and that each natural number has a 'next' natural number. The idea is to think of the proof as a set of individual dominoes standing in a line, so that if one is pushed over then it topples the next one, and so on, down the line. To topple all the dominoes we then only need to push the first one over. Suppose there is a statement $P(n)$ that we want to prove for each natural number n. First prove that it is true when $n = 1$. Then prove that if it is true for $n = k$, then it will also be true for $n = k + 1$. Since it is true for $P(1)$ we then know it is true for $P(2)$. But this means it is true for $P(3)$, and so on, proving truth for all natural numbers.

9 An example of proof by induction

In a temple in Hanoi there are three poles and a stack of 64 golden discs in increasing order of size. The priests must move all the discs to a new pole, with the rule that a smaller disc can only be placed on top of a larger disc. According to the legend, the world will end when the task is completed. How long will it take? This puzzle, called the Tower of Hanoi, can be solved using proof by induction, conjecturing that when there are n discs the puzzle takes $2^n - 1$ moves to finish. When $n = 1$ the puzzle takes 1 move, which fits the formula. Suppose the formula is true for k discs. If we had $k + 1$ discs, we would first move the top k of them to the second pole (taking $2^k - 1$ moves, by the formula), then move the biggest one to the last pole (1 move), then move the k discs to the last pole ($2^k - 1$ moves) for a total of $2(2^k - 1) + 1 = 2^{k+1} - 1$ moves. So the formula is true by induction, and the priests need $2^{64} - 1$ moves (which will take longer than the age of the universe).

10 Different orders of logic

Proof by induction not only uses the properties of natural numbers, but in some sense it *defines* the natural numbers. The fact that any natural number has a unique successor is called the axiom of induction in the Peano axioms that define arithmetic. People encountering proof by induction for the first time are often sceptical of it, as it is far less obvious than the other methods of proof. Logicians are wary of it because it is a proof technique using second-order logic, while the other methods use first-order logic. First-order logic makes statements about variables, but second-order logic makes statements about first-order statements. Most axioms of mathematics try to stick to first-order logic, but infinity only really makes sense with second-order logic.

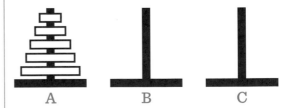

A   B   C

TALK LIKE A GENIUS

❦ A subtly incorrect version of proof by induction can show that all horses are the same colour. The result is clearly true for one horse. Suppose that the horses in any collection of n horses are the same colour and consider a collection of $n + 1$ horses. Remove one; the remaining n horses are the same colour as each other. Put it back and remove a different horse. The horses in this second collection of n horses are also the same colour as each other. But the two collections overlap, so must all be the same colour. Hence, all $n + 1$ horses are the same colour, meaning the proof is true by induction. ❧

❦ Mathematical proofs traditionally end with an empty □ or filled-in square ■ as an indication that the proof is finished. They may also end with QED, standing for *"quod erat demonstrandum"*, meaning "what was to be shown". Proofs by contradiction often use a different symbol, such as a lightning bolt ↯ or double arrows →←. ❧

❦ The Dutch mathematician Brouwer is probably most famous in mathematics for proving his fixed-point theorem in topology. For example, when you stir your coffee, Brouwer's result proves that there will be at least one molecule in the same place after the stirring as before. But his proof does not explain how to find which molecule it is – it is a non-constructive proof. Ironically, Brouwer later became one of the biggest advocates for a constructionist philosophy. ❧

❶ FALSE – A proof is an irrefutable sequence of logical steps, starting from known truths, that deduces the statement in question is true.

❷ FALSE – The negation is 'There is a sheep that does not have four legs'.

❸ TRUE – It says that either 'A' is true or 'not A' is true.

❹ FALSE – The contrapositive is 'if I have not had alcohol then I am not drunk', and is logically equivalent to the original statement.

❺ TRUE – Induction first proves a statement true for $n = 1$, then shows that if it is true for $n = k$ then it is also true for $n = k + 1$, with these steps together showing it is true for all natural numbers.

THE BLUFFER'S SUMMARY

A mathematical proof starts from a set of truths and uses logic to deduce new truths.

Sizes of infinity

'The essence of mathematics lies in its freedom.'

GEORG CANTOR

In possibly the most mind-blowing theorem of mathematics, Georg Cantor proved that infinity comes in different sizes. His results were dismissed as nonsense by contemporaries in both mathematics and theology, who felt that they threatened the very idea of God. Dying alone and in poverty in a sanatorium in 1918, filled with paranoia and depression, Cantor is often said to have been driven mad by the contradictory nature of his results. Yet today, his ideas about sets and cardinality form the foundation of mathematics, and have thrown up questions, the answers to which are stranger than anyone could ever have expected.

How is it possible that one infinity could be bigger than another – and does it even make sense to ask the question of how big infinity is?

1 The infinity symbol ∞ was invented by the ancient Greeks.
TRUE / FALSE

2 The set of even numbers is the same size as the set of all whole numbers.
TRUE / FALSE

3 The set of all integer fractions is bigger than the set of all whole numbers.
TRUE / FALSE

4 Mathematicians have proved that there is no infinity bigger than that of the whole numbers but smaller than that of the decimal numbers.
TRUE / FALSE

5 There is no biggest infinity.
TRUE / FALSE

1 What a set is

A set is a collection of things. Examples include the set of all cars (which is finite in size), the set of all one-legged sheep (which is empty) and the set of all whole numbers (which is infinite). Mathematicians use curly brackets to mean a set, so {1} is the set containing the element '1'. To compare the sizes of two sets, we do not need to count how many things are in each. A waiter with a set of of knives and a set of forks will simply move around the room laying down one knife beside each fork, and if none are left over at the end then the two sets were the same size. The size of a set is called its cardinality, and the German mathematician Georg Cantor was the first person to investigate sets of infinite cardinality.

2 The paradoxical nature of infinity

'The whole is greater than part'. So said Euclid, but Cantor showed that such 'obvious' results are no longer true for infinite sets. For example, the set of all even numbers {2, 4, 6, 8, ...} is a part of the set of all positive integers (known as the natural numbers) {1, 2, 3, 4, 5, ...}. But taking each natural number n and pairing it with the even number $2n$ (so 1 is paired with 2, 2 with 4, 3 with 6, and so on) we see that no number from either set remains unpaired with another, and so the two collections must be the same size. Any set that has the same size as the natural numbers is said to be countable.

3 How big the set of fractions is

One attempt at discovering a 'bigger' infinity than that of the natural numbers was to consider the set of all integer fractions, known as the rational numbers. This includes numbers like $\frac{1}{2}$ and $\frac{21}{3}$. The cardinality of this set can be thought of as 'infinity squared', because we can choose an infinity of different numbers to become the numerator, and another infinity of numbers to become the denominator. But Cantor found a way to systematically 'count' the fractions (see diagram at upper right) in such a way that none were left out, showing that it was possible to pair up every fraction with a natural number. So the rational numbers are countably infinite. Perhaps it was the case that all infinite sets were actually the same size after all.

4 Whether an infinity can be uncountable

Decimal numbers have an infinite sequence of digits after the decimal point. Even a number like 0.5 can be thought of as 0.50000 ... with an infinite sequence of zeros. Their cardinality can be thought of as '10 to the power of infinity' because we have ten choices of digit for each of the infinite decimal places. Cantor used a wonderfully clever proof called the 'diagonal argument' to show that this collection of all decimals could never be paired up with the natural numbers in such a way that no decimal was left over. So the cardinality of the decimal numbers must be bigger than that of the natural numbers. This was the first uncountable infinity ever discovered.

5 How Cantor's diagonal argument works

Cantor's diagonal argument is an example of a method of proof called 'proof by contradiction'. It starts by assuming that every decimal can be paired with a natural number in such a way that none are left over. This means we can list the decimals, one by one. Now, using this list, we create a new number by the following method. Choose a digit for the first decimal place of the new number that is different to the digit in the first decimal place of number (1). Then choose a digit for the second decimal place of the new number that is different to the digit in the second decimal place of number (2). Similarly with the third digit of number (3), the fourth digit of number (4) and so on. The new number has been constructed to be different to every number in the list, meaning it is not in the list, and so it contradicts the assumption that it was possible to list every decimal.

(1) 0.**1**00000...

(2) 0.1**2**02759...

(3) 0.14**6**232...

(4) 0.223**5**60...

➜ **0.2376...**

6 **How to create ever bigger infinities**
As if finding two different sizes of infinity were not mind-blowing enough, Cantor went on to show that there is an infinite hierarchy of infinities. He did this by using the idea of a 'power set': a new set consisting of all possible collections of elements from a given set. For example, if we take the set {1,2,3} (which has cardinality 3) its power set contains the empty set, {1}, {2}, {3}, {1,2}, {1,3}, {2,3}, and {1,2,3}, and so has cardinality 8. The formula is that, if the original set has size n, then its power set has size 2^n, which is always bigger than n. Amazingly, the same conclusion is true for infinite sets: the power set of an infinite set creates a new set of a bigger infinity.

7 **How Russell's paradox undermined set theory**
One reason that Cantor's work was criticized was because of logical paradoxes deep in the heart of set theory. The most serious of these was Bertrand Russell's paradox, which showed that not everything could be a set. The paradox starts by noticing that there are two types of set: those that contain themselves and those that do not. For example, the set of all things that are not cats is itself not a cat. Now consider the set of all sets that do not contain themselves. Call it S. Does S contain itself? If it does, then, by definition, it does not, but if it does not, then it should. Russell's paradox has since been resolved in a number of different ways, each of which restricts the kinds of objects that are allowed to be called sets.

8 **What the continuum hypothesis says**
Mathematicians use the symbol $\aleph_0$ (pronounced aleph-zero) to represent the cardinality of the natural numbers. The next biggest cardinal number is called $\aleph_1$, then $\aleph_2$ and so on. It is an obvious question to ask whether the cardinality of the decimal numbers is $\aleph_1$. That is, is the set of decimals the smallest uncountable set? The continuum hypothesis, put forward by Cantor in 1878, is that the answer to this question is yes. In 1900, David Hilbert (one of Cantor's few supporters) put the question first on his list of 23 unsolved problems in mathematics.

9 **How the continuum hypothesis was resolved**
The continuum hypothesis had to wait until 1963 to be fully resolved, and its answer was something that nobody, least of all Cantor, had expected. Kurt Gödel and Paul Cohen showed that the continuum hypothesis could neither be proved nor disproved using the standard axioms of mathematics. That is, assuming it to be true would never contradict any other theorem in mathematics, but neither would assuming it to be false. As such it was one of the first practical demonstrations of Gödel's incompleteness theorem (see page 56).

10 **What the generalized continuum hypothesis is**
If a set has cardinality $\aleph_a$ then does its power set have cardinality $\aleph_{a+1}$? The generalized continuum hypothesis states that the answer is 'yes', and, like the continuum hypothesis, it cannot be proved to be either true or false. This does not stop mathematicians from having an opinion on whether it is true or false, with many believing that a more sophisticated model of set theory will resolve the question one way or another. To throw in a final piece of intrigue, the comparison of cardinalities of sets is only possible using the axiom of choice, another hypothesis that can neither be proved nor disproved under our axioms of mathematics. It may be possible that there are multiple sets whose cardinalities are bigger than $\aleph_0$, but whose sizes are not comparable with each other, meaning that the phrase 'next biggest cardinal' makes no sense.

TALK LIKE A GENIUS

❝ Theologians were concerned that Cantor's new hierarchy of infinities threatened the absolute infinity that was associated with the Christian god. Cantor was a devout Christian and even wrote to Pope Leo XIII to explain his ideas and to show that they could exist harmoniously with his faith. ❞

❝ While it is commonly stated that Cantor went mad because of the paradoxical nature of his ideas and his inability to resolve the continuum hypothesis, it is believed now that he may have had bipolar disorder. The intense criticism of his work by men such as Leopold Kronecker resulted in Cantor leaving mathematics for a time and working in arcane areas of research, such as trying to prove that Francis Bacon wrote Shakespeare's plays, or that Joseph of Arimathea fathered Christ. ❞

❝ A more common version of Russell's paradox is that of the barber's paradox. In a town there is a barber who shaves every man who does not shave himself. But who shaves the barber? ❞

WERE YOU A GENIUS?

1 FALSE – It was first introduced by John Wallis in 1655, possibly based on the last letter of the Greek alphabet, ω, or on a Roman numeral for 1000, CIƆ

2 TRUE – By pairing each even number $2n$ with the whole number n, we find a correspondence between the sets that leaves none out on either side, so they must be the same size.

3 FALSE – It is possible to write out all the fractions in a numbered list so that none are left out.

4 FALSE – Mathematicians have proved that it is undecidable whether such a size of infinity can exist.

5 TRUE – Cantor used power sets to construct an infinite hierarchy of infinities, each bigger than the last.

THE BLUFFER'S SUMMARY

Infinity comes in different sizes and there is no such thing as a 'biggest' infinity.

Gödel's incompleteness theorem

'We must know. We shall know!'

DAVID HILBERT

In 1900, at the International Congress of Mathematics, mathematician David Hilbert set out 23 problems that would influence the direction of mathematics for the next century. Number 2 on his list was the following: 'Show that arithmetic is free of contradictions.' After the problems with Cantor's set theory, and the paradoxes it contained, mathematicians wanted to ensure that such paradoxes were not lurking deeper in the heart of their subject. Gödel's incompleteness theorem was an attempt to answer this question, but far from confirming that 'We shall know', it brought Hilbert's programme crashing down by saying that 'We can never know'.

Is it possible to know everything? Does every question have an answer? No, says mathematics, in a cruel twist of logic.

1 All mathematical statements can be proven to be either true or false.

TRUE / FALSE

2 Nobody has ever found a contradiction in our currently accepted axioms of mathematics.

TRUE / FALSE

3 If 0 = 1, then all sheep have six legs.

TRUE / FALSE

4 This sentence is false.

TRUE / FALSE

5 We have never found a mathematical statement incapable of being proven either true or false.

TRUE / FALSE

TEN THINGS A GENIUS KNOWS

1 **What an axiom is in mathematics**
An axiom is a statement that is so obvious that it can be assumed to be true without requiring proof. For arithmetic, our axioms include things like: 'All numbers are equal to themselves', '0 is a number', and 'For any whole number, there is a next whole number'. Logic deals with combining statements, and how the truth of a new statement follows from those of which it is formed. For example, we may have the two statements 'I am in England' and 'It is raining'. We can combine these to make a new statement: 'If I am in England then it is raining'. Logic tells us that this new statement is false if it happens that we are in England and it is not raining.

2 **How to choose a set of axioms**
Mathematicians have long debated what the 'correct' axioms of arithmetic should be. Ideally, we want to choose the shortest possible list of axioms that allows us to do all the mathematics we want, and we want to choose them in such a way that no contradictions can occur. A contradiction happens when a statement is both true and false at the same time. If this happens, then classical logic tells us that all other statements can be inferred from it. So, if 0 = 1, then Santa Claus exists.

3 **The problem with paradoxes in set theory**
Set theory, as invented by Cantor, looked to be a very promising way to define what numbers were. A set is just a collection of items. The number 0 can be equated with the empty set, {}, which contains nothing. The number 1 can be equated with the set containing 0, {0}, since this has a single element. The number 2 can be equated with the set containing 0 and 1, {0,1}, as this has two elements, and so on, building up the natural numbers sequentially. The problem with this is that naïve set theory runs into paradoxes. Cantor proved that, given any set, we can always construct a bigger one. But he also showed that the set of all sets was the biggest one possible. Both of these statements cannot be true. Are there axioms for set theory, and thus arithmetic, that are free of such contradictions?

4 **What consistency and completeness are**
A set of axioms is called consistent if it never results in a contradiction. A set of axioms is called complete if every statement resulting from those axioms can be either proved or disproved. David Hilbert wished to begin a programme of study that would not only find a set of axioms for arithmetic that were both complete and consistent, but that would also create an algorithm that could decide the truth or falsehood of any given statement. In 1931, the Austrian mathematician Kurt Gödel, at the age of 25, addressed Hilbert's programme with two devastating results.

5 **What Gödel's first incompleteness theorem says**
This theorem states that any consistent axioms of mathematics that are sufficiently complex to encode arithmetic will be incomplete. Gödel proved his theorem by creating a way to talk about provability using only the language of arithmetic. Then any system – let us call it F – complicated enough to make statements about numbers could also make statements about its own provability. Gödel's genius masterstroke was then to write the sentence 'This sentence cannot be proven within the system F', using the language of the system F. If this statement were provable, then it contradicts itself, meaning that the system is inconsistent. So if the system F is consistent, then the statement must be unprovable (and therefore true!), meaning that F is incomplete.

6 **What Gödel's second incompleteness theorem says**
In his first incompleteness theorem, Gödel showed that any consistent axioms of arithmetic would contain theorems that were true but unprovable. In his second incompleteness theorem, Gödel showed that it was never possible to demonstrate the consistency of a set of axioms from within the system itself. This was the final blow to Hilbert's programme. Gödel had first said that mathematics could not be complete, and now he had said that we could never prove it was consistent either.

7 **Consequences of the incompleteness theorems**
Although a set of axioms cannot be proven to be consistent within themselves, we can build bigger systems within which the smaller systems are complete. Mathematicians Zermelo and Fraenkel designed a form of set theory that is now taken to

be a standard foundation of mathematics. It can be proven to be consistent by assuming the existence of something called an 'inaccessible cardinal'. Yet, because it is consistent, mathematicians must live with the fact that it is incomplete and that some statements it makes will be unprovable.

8 Whether there are unprovable statements in mathematics

Despite knowing that unprovable statements existed in mathematics, people expected such statements to be so obscure that they would never appear in everyday mathematics. It therefore came as a great shock when Gödel went on to prove that the continuum hypothesis, which stated that there was no size of infinity in between the natural numbers and the real numbers, was unprovable. That is, it could be taken to be true or false, and no contradictions would arise either way. Another example of an unprovable statement is called the well-ordering theorem. This says that if you take any set of objects, there is a way of ordering them so that in any non-empty subset there will be a least element. The usual ordering 'less than or equal to' ($\leq$) works fine for the positive whole numbers, but not for all of the integers since 'the set of numbers less than 0' would have no least element. The well-ordering theorem feels like it should be false, yet it is unprovable.

9 What the axiom of choice is

In a collection of pairs of shoes, we can choose one from each pair by specifying that we always pick the left shoe. But with a collection of socks, where there is no difference between the two socks in a pair, there is no algorithmic way to decide which one to pick. A similar problem arises in mathematics:

given a (potentially infinite) collection of sets, is it possible to choose one element from each set? The axiom of choice says that the answer is yes, but it turns out to be another unprovable statement. Its conclusion appears so obvious that most mathematicians are happy to accept it as an axiom. However, it results in some strange paradoxes, such as the Banach-Tarski paradox. This says that it is possible to break up a ball into finitely many pieces and reconstruct them so as to have two balls, each the same size as the first. Mathematicians have also shown that the axiom of choice, which is intuitively true, is logically equivalent to the well-ordering theorem, which is intuitively false.

10 Whether Hilbert's question has been resolved

There remains no consensus on whether Hilbert's question about the consistency of arithmetic has been resolved, despite Gödel's theorems. Although Gödel's results show that arithmetic cannot be proven consistent within itself, this does not mean that a meta-mathematical proof using ideas outside the system might not be accepted by the mathematical community. The majority of mathematicians do not worry either way, content to believe that the Zermelo-Fraenkel axioms will never be shown to be inconsistent.

TALK LIKE A GENIUS

❧ In 1930, at a conference in Königsberg, David Hilbert gave his retirement address. He argued his belief that all mathematical problems could be solved. He said "For the mathematician there is no Ignorabimus . . . The true reason why [no one] has succeeded in finding an unsolvable problem is, in my opinion, that there is no unsolvable problem. In contrast to the foolish Ignorabimus, our credo avers: We must know. We shall know!" Two days later, at the same conference, Gödel announced his results. ❧

❧ The philosopher Wittgenstein criticized Gödel's work, leading Gödel to say "Has Wittgenstein lost his mind? He intentionally utters trivially nonsensical statements." Logicians continue to argue over whether Wittgenstein truly understood Gödel's work. ❧

❧ Gödel's incompleteness theorems are often misapplied to other fields, such as quantum physics. But results like Heisenberg's uncertainty principle, in which we cannot know both the position and momentum of a particle, have nothing to do with the unknowability of statements in mathematics. ❧

❧ To test whether your guests are mathematical geniuses, try the following joke. What is an anagram of Banach-Tarski? Banach-Tarski Banach-Tarski! ❧

THE BLUFFER'S SUMMARY

Statements will always exist in mathematics that cannot be proven to be either true or false.

Turing machines

'It is possible to invent a single machine which can be used to compute any computable sequence.'

ALAN TURING

When we pick up our smartphones, we take it for granted that they can make calls, play games, calculate sums and find our location. But in 1936, when Turing wrote the sentence above, the idea of a 'universal computer' that could perform many different tasks seemed far fetched. The concept of a Turing machine not only paved the way for modern concepts of computer software, but solved a deep mystery in mathematical logic about whether machines could decide the truth of mathematical theorems. Computers have come a long way since 1936, but Turing's work shows us that there will always be limits to what they can achieve.

Turing's genius has left its mark in many ways on the world, not least in leaving us questioning what intelligence really is.

1 An algorithm is a finite list of instructions that a computer follows to produce an output.
TRUE / FALSE

2 The input to a computer program can be another computer program.
TRUE / FALSE

3 A Turing machine is at least as powerful as any modern computer.
TRUE / FALSE

4 Software exists that can tell if a computer program will stop or whether it will run forever.
TRUE / FALSE

5 To pass a Turing test, a computer must answer a series of questions and convince a jury that it is a human.
TRUE / FALSE

TEN THINGS A GENIUS KNOWS

1 What the *Entscheidungsproblem* is

The 17th-century genius Gottfried Leibniz was one of the first to build a mechanical calculating device, called the Stepped Reckoner. After doing so he asked the question of whether computers would one day be able to prove theorems in mathematics. This question was re-posed by David Hilbert in 1928 and called the *Entscheidungsproblem*, or 'Decision Problem'. More specifically, the Decision Problem asked if there could exist an algorithm that would take as input a mathematical statement and output 'yes' or 'no', according to whether the statement was a true conclusion from the axioms of mathematics. The problem gained special significance a few years later when Kurt Gödel showed that some statements in mathematics were unprovable.

2 What an algorithm is

To be able to answer the *Entscheidungsproblem*, mathematicians first had to make precise what they meant by an 'algorithm'. This meant encapsulating the idea that a mathematical statement or formula was 'calculable' if a human could sit down with a pencil and paper and eventually produce an answer. Turing machines were one way to formalize this idea: that a machine would follow a finite set of instructions, manipulating a set of symbols to produce an output. An American mathematician, Alonzo Church, came up with an equivalent answer to the problem using a method called lambda-calculus. Both Turing and Church are credited with solving the *Entscheidungsproblem*, and both types of solution led to major advances in computer science.

3 How a Turing machine works

A Turing machine is the simplest possible device for performing computations. It consists of a length of tape that is divided into cells, with each cell containing one of a fixed number of symbols. The tape extends infinitely to the left and the right, but only one special symbol, called the blank, is allowed to appear infinitely often on the tape. A device called the head can read the symbols, rewrite the symbols, and move the tape one space left or right. The machine also has a 'state register', which stores its current state. For the machine to operate, it must be given an input and a finite table of instructions, which tells the head what to do when it is in a particular state and sees a particular symbol.

4 How powerful a Turing machine is

The Turing machine was designed to be the simplest possible device for doing computations, yet mathematicians have shown that it is just as powerful as any modern computer in existence today, in the sense that it can do anything a modern computer can do. We might think that by adding extra things in to the design we could expand the types of computations it could do; for example, if we had multiple tapes, or could move extra spaces left or right, or could choose not to move at all. But none of these additions change the power of the machine, even though they may allow it to do computations faster or more efficiently. The key is that the tape in the Turing machine is infinitely long, which essentially means that it has an unlimited amount of memory.

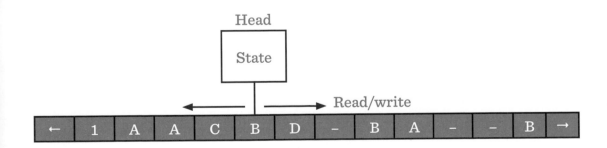

5 The importance of a universal Turing machine

One of Turing's genius ideas was that one Turing machine could be used to simulate the work of another Turing machine. The input to this 'universal Turing machine' would be the instructions for another Turing machine, together with its input, and it would produce the same output as that machine. This means that programs can run other programs, and the idea paved the way for computers that could store programs in the same memory used for storing data. Before this idea, computers were either designed for a single purpose, such as Leibniz' Stepped Reckoner, or were programmed using punch cards. It was not until 1948 that a computer was made that could store a program in its memory.

6 What the halting problem is

Most of us have been in a situation in which our computer appears to freeze, leaving us unsure whether to wait for the process to stop or to give up and restart the machine. This problem was anticipated before modern computers were invented, by Church and Turing's work in algorithms in the 1930s. In a reformulation of Hilbert's *Entscheidungsproblem*, the two men asked whether there was an algorithm that could decide whether a given computer program would halt or whether it would continue to run forever. This is known as the halting problem.

7 How the halting problem solved the *Entscheidungsproblem*

Turing's great result was to prove that the halting problem was undecidable. That is, he showed that there could never exist a single algorithm that could decide whether programs would stop. For any *particular* program we might be able to figure out if it would stop or not, but there would never be a one-size-fits-all method of deciding the answer for all programs. It turns out that Hilbert's question about whether mathematical statements could be proven true can be translated into exactly this question about whether programs will halt. For example, one current unsolved problem is the Goldbach conjecture, which claims that every even number is the sum of two primes. We could write a program that checks the result for every even number, and halts if it finds a counter-example. The statement that the Goldbach conjecture is true is then equivalent to this program never halting, so if a program could predict the halting of the program it could also prove the Goldbach conjecture true or false.

8 Whether the human brain is a Turing machine

Turing's answer to the halting problem showed that there were some things Turing machines could not do. But what if there were machines more powerful than a Turing computer? In particular, could there be physical processes, such as the workings of a human brain, that cannot be simulated by a Turing machine? The answer to this is an open question. Turing himself investigated the possibilities of 'hypercomputers', but showed that there would always remain some questions that were undecidable.

9 What the Turing test is

Another way of asking whether a Turing machine is as good as a human brain is to ask whether a computer could ever pass as a human. The questions of whether machines can really think, or what it means to be intelligent, have long been discussed by scientists and philosophers alike, but Turing brought the discussion back to earth with the design of a very simple test. A computer would be said to have passed the 'Turing test' if it can convince a significant proportion of a jury, through answering a series of questions posed to it, that it is really a human.

10 How to win money with the Turing test

The Loebner prize was set up in 1991 as an annual competition based on the Turing test. The exact rules have changed from year to year, but the general format is that an interrogator has a simultaneous conversation with both a computer and a real human, and must decide which is which. Prizes are given each year for the most-human ChatterBot, and there are two one-time-only prizes. There is $25,000 up for grabs for the first program that can convince the judges that the human is the computer, and $100,000 available for the first program that can pass a Turing test that includes questions that are auditory, visual and textual. So far the big prizes have not been won, and Turing's prediction that we would have artificial intelligence by the year 2000 has not come true.

TALK LIKE A GENIUS

❧ The Turing test is also known as the "Imitation Game", which is the title used in the 2014 film about Alan Turing's work in cracking the German Enigma code during the Second World War. ❧

❧ The Enigma machines used by the Germans during the Second World War were the most difficult ciphers ever seen. They were designed in such a way that the message being sent became part of the encryption process itself. With trillions of possible settings there was no way messages could be deciphered by hand. Turing helped design and build a computer called the Bombe, which used logic to eliminate incorrect Enigma settings, allowing the Allies to read German messages. Historians estimate that Turing's efforts shortened the war by at least two years, saving millions of lives. ❧

❧ In our modern online world, the idea of a Turing test is more important than ever, both in distinguishing who is a robot and who is a human. ChatterBots can be so convincing that many people fall victim to scams asking for them to reveal personal details or send money. On the other side, websites often ask users to prove they are human by passing tests called CAPTCHA, like typing in words from a distorted image. CAPTCHA actually stands for Completely Automated Public Turing test to tell Computers and Humans Apart. ❧

WERE YOU A GENIUS?

❙ TRUE – The idea is modelled on human 'computers' who solved mathematical problems using pencils and paper.

❷ TRUE – This idea is the basis for software on computers, and the idea that programs can be stored in the same memory as that used for data.

❸ TRUE – A Turing machine can perform any computation that a modern computer can perform.

❹ FALSE – Turing proved that no such software could exist by showing that the halting problem was undecidable.

❺ TRUE – Questions are asked simultaneously to a computer and a human, and the computer passes the test if the jury erroneously believes that it is the human.

THE BLUFFER'S SUMMARY

A Turing machine is a device that can simulate any computer. But problems will always exist that a Turing machine cannot solve, giving limits to what computers can do.

Platonic solids

'Geometry will draw the soul towards truth.'

PLATO

The five Platonic solids are often considered the most beautiful and perfect of all the shapes. They were known long before the philosopher Plato, but he popularized them by equating them with four classical elements (earth, air, fire and water), saying that the fifth solid, the dodecahedron, was used by the gods for arranging the constellations in the heavens. Subsequent mathematicians have remained fascinated by these objects, uncovering deeper truths than even Plato had imagined.

Only five shapes in three dimensions have perfect symmetry – not because nobody has been genius enough to discover more, but because of the intrinsic structure of geometry.

ARE YOU A GENIUS

1 The icosahedron has 20 pentagonal sides.
TRUE / FALSE

2 The pyramids of Giza in Egypt are tetrahedral.
TRUE / FALSE

3 A football is a Platonic solid, as it is built of regular hexagons and pentagons.
TRUE / FALSE

4 Five simultaneous tetrahedra can be drawn between the vertices of a dodecahedron.
TRUE / FALSE

5 There are 24 ways of rotating a cube to leave it looking the same.
TRUE / FALSE

TEN THINGS A GENIUS KNOWS

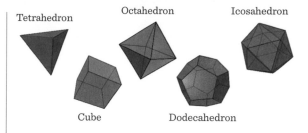

Tetrahedron Octahedron Icosahedron

Cube Dodecahedron

1 **What a polyhedron is**
A polyhedron (plural: polyhedra) is a three-dimensional (3-D) shape made up of flat sides (called faces) and straight edges. The pyramids of Giza are polyhedra, as they are made up of four triangular faces plus a square base. A cone is not a polyhedron, because its base is circular and hence not made up of straight sides. Mathematicians like to think about a special class of polyhedra: those that are convex. In a convex polyhedron, any line joining two points in the shape must itself stay within the shape. The pictured star is a polyhedron because it is made up of straight sides, but it is not convex because the line drawn joining two arms of the star goes outside the shape.

2 **What a Platonic solid is**
A Platonic solid is a convex polyhedron in which every angle is the same size, every edge is the same length and every face is the same shape. In addition, there are the same number of faces meeting at each corner (known as a vertex). These conditions mean that every face must be a regular polygon, such as an equilateral triangle, a square, or a regular pentagon. In fact, it turns out that these three shapes are the only polygons that can be faces of Platonic solids. This is because, at each vertex of the polyhedron, there must be three or more of the faces meeting. Polygons with more than six sides have angles that are too big to fit three or more around a vertex. Using hexagons results in a flat shape rather than a 3-D shape.

3 **Which shapes are Platonic solids**
Using equilateral triangles, we can build three different Platonic solids: the tetrahedron (four triangles), the octahedron (eight triangles) and the icosahedron (20 triangles). Using squares we can build another Platonic solid: the humble cube, also called a hexahedron (six squares), and using pentagons we can make one final Platonic solid: a shape called a dodecahedron (12 pentagons). No other convex polyhedra can be made using only equilateral triangles, squares or pentagons.

4 **The dualities of Platonic solids**
The cube has six faces and eight vertices, while the octahedron has eight faces and six vertices. This is not a coincidence, and hints at deeper relationships between these beautiful shapes. Take a cube and start filing down the corners, leaving flat surfaces in their place. Do this for all the corners simultaneously and continue until these new faces meet each other. The result will be an octahedron. Starting instead with an octahedron and repeating the procedure will create a cube. We say that the cube and the octahedron are dual to each other. A similar duality exists between the dodecahedron (12 faces and 20 vertices) and the icosahedron (20 faces and 12 vertices). (One of the intermediate steps in carving a dodecahedron into an icosahedron produces a football, which is made of pentagons and hexagons.) Since there are only five Platonic solids, what is the tetrahedron (four faces and four edges) dual with? The answer: the tetrahedron is dual with itself.

5 **How Platonic solids fit inside one another**
The various Platonic solids fit inside one another in many surprising ways. A tetrahedron has six edges, and these can be arranged so as to lie along a diagonal of each of the six faces of a cube. Taking the opposite diagonal of each face results in a second tetrahedron inside a cube. Together, these two tetrahedra form a stellated octahedron: that is, an octahedron with a pyramid on each face. We can also arrange a

tetrahedron so that its vertices coincide with the vertices of a dodecahedron. Since the dodecahedron has 20 vertices and the tetrahedron has four, we can fit five tetrahedra simultaneously inside a dodecahedron. We can similarly fit a cube inside a dodecahedron with the vertices matching, and there are five ways of doing this.

6 How many symmetries the Platonic solids have

The Platonic solids have an exceptionally high number of symmetries and these symmetries form a mathematical group (see page 117). A symmetry here is any action one can do to a shape that leaves it looking unchanged. For a cube (and, by duality, an octahedron) there are 24 different rotations that leave it unchanged. If reflections are also allowed this brings us up to 48. There are 12 tetrahedral rotational symmetries, and 60 dodecahedral/ icosahedral rotational symmetries. Can you find them all?

7 What Euler's formula says

The Swiss mathematician Leonhard Euler was the first to notice the following surprising fact. If you take any of the Platonic solids and add up the number of vertices, subtract the number of edges, and add on the number of faces, the answer is 2. For example, the cube has 8 vertices (corners), 12 edges and 6 faces, and $8 - 12 + 6 = 2$. This became known as Euler's polyhedral formula, and is usually written $V - E + F = 2$. Mathematicians later found that this formula was true of *any* convex polyhedron, and holds because all of these shapes are topologically equivalent to spheres. (See page 89.)

8 Whether Platonic solids exist in higher dimensions

In the same way that we can use regular polygons to build symmetric 3-D shapes, we can use the five Platonic solids to build symmetric 4-D shapes. It turns out that in four dimensions there are six Platonic solids (or 'regular convex polytopes'). Combining these, we can construct yet higher dimensional shapes, and we might expect the number of such shapes to increase with each dimension. Yet it is in dimension four that we find the most Platonic solids: in every dimension above four there are only three Platonic solids,

which are the analogues of the tetrahedron, cube and octahedron.

9 Where to find Platonic solids in nature

Platonic solids may be found naturally in both animate and inanimate objects. Crystal structures in minerals will mirror the configuration of the molecules, which is frequently highly symmetric. For example, magnetite is a type of iron oxide, and its iron atoms like to form six-fold symmetric configurations around oxygen atoms, leading to octahedral crystals. Pyrite can form perfect cubes because of the arrangements of iron and sulphur atoms. In biology, viruses are found with Platonic solid structures, because the symmetry means that the shapes have very simple manufacturing instructions, saving space in the virus's genome. Herpes and HIV both have icosahedral-shaped shells.

10 What shapes go beyond Platonic solids

If we relax the constraint that Platonic solids must be convex, we may add four more shapes to our list. These are called the Kepler–Poinsot polyhedra and are related to the dodecahedron and icosahedron. If, instead, we no longer care that the faces must all be the same shape, we obtain the 13 Archimedean solids. In these shapes all the faces are regular polygons, and each vertex has exactly the same set of shapes meeting there. The football, made of regular hexagons and pentagons, is an Archimedean solid. Taking the duals of the Archimedean solids gives us the Catalan solids: all their faces are identical, but these faces are no longer regular polygons.

TALK LIKE A GENIUS

❝ The great astronomer Johannes Kepler believed for a long time that the Platonic solids accounted for the distances between the planets. His model of the solar system placed the planets (Mercury, Venus, Earth, Mars, Jupiter and Saturn) in concentric Platonic solids, starting with the octahedron in the centre, followed by the icosahedron, dodecahedron, tetrahedron, and the cube on the outside. This beautiful theory not only explained the placement of the planets in the solar system, but explained why there were only six planets. Although his theory was wrong, the detailed calculations needed to come up with it eventually led to Kepler's laws of orbital dynamics and his realization that planets travelled in ellipses, a discovery that would forever change astronomy. ❞

❝ Although the Platonic solids are named after the Greek philosopher Plato, they were known to humanity long before him. Examples exist of dice with Platonic solid symmetries carved by Neolithic man, and there is evidence that the Egyptians and the Pythagoreans knew of the Platonic solids. However, it was not until Euclid that we saw the first proof that only five Platonic solids existed. ❞

❝ A four-dimensional cube is called a tesseract and is built from eight 3-D cubes. La Grande Arche de la Défense near Paris is a model of such a 4-D cube.❞

WERE YOU A GENIUS?

1 FALSE – It has 20 sides that are each equilateral triangles.

2 FALSE – The pyramids of Giza consist of four triangles for their sides plus a square base, while tetrahedra are made of three triangular sides plus a triangular base.

3 FALSE – Platonic solids must have all their faces identical. Footballs are examples of Archimedean solids, which look identical around each vertex.

4 TRUE – The dodecahedron has 20 vertices and the tetrahedron has four, so five tetrahedra can sit inside a dodecahedron.

5 TRUE – There are 24 rotational symmetries of a cube.

THE BLUFFER'S SUMMARY

There are exactly five 3-D shapes made of straight sides in which every edge is the same length, every angle is the same size and every face is the same shape.

Non-Euclidean geometry

'You must not attempt this approach to parallels... I have traversed this bottomless night, which extinguished all light and joy in my life.'

FARKAS BOLYAI

Euclidean geometry, the geometry we all learn in school, is built from just five axioms, or postulates – five truths so elementary that they do not need proof to be believed. However, the fifth postulate (also called the parallel postulate) is significantly more complex than the others, making people wonder if it could be proved from the other four. Mathematicians struggled for 2000 years to find this elusive proof, with the bombshell finally falling in the 19th century that such a proof was impossible. The new geometry this created has gone on to explain the shape of many things in the natural world, such as corals, mushrooms and leaves.

By questioning the very foundations of geometry, mathematicians created whole new worlds that remarkably echoed the shapes of nature.

1 On a piece of paper, we can draw infinitely many lines parallel to the edge of the paper, all going through the centre of the paper.

TRUE / FALSE

2 Triangles drawn on the surface of the Earth have angles that sum to more than 180°.

TRUE / FALSE

3 Two lines are called parallel if they always remain the same distance apart.

TRUE / FALSE

4 On a sphere, there is no such thing as parallel lines.

TRUE / FALSE

5 Many plants and corals have hyperbolic geometry because it maximizes their surface area for collecting nutrients.

TRUE / FALSE

TEN THINGS A GENIUS KNOWS

1 The importance of Euclid

Euclidean geometry is named after the ancient Greek mathematician Euclid, whose work *The Elements* was the standard textbook on geometry right up to the 20th century. Some of the results in this work were known before Euclid, but he was the first to write mathematics as we recognize it today, starting with axioms and applying logical deductions to arrive at new theorems. This idea of 'proof' became the foundation of the mathematical method, and Euclid's results on geometry and number theory were the unquestioned foundation of mathematics itself.

2 Euclid's axioms of geometry

Euclid aimed to start his study of geometry with the smallest possible number of axioms, or postulates, and he came up with a list of five upon which all geometry would be based. (In his axioms, 'straight line' means the shortest distance between two points.) First, a straight line can be drawn between any two points. Second, any finite line segment can be extended to an infinite straight line. Third, a circle can be drawn with any centre and any diameter. Fourth, all right angles are equal to one another. And fifth, if a straight line falling on two straight lines makes the interior angles on the same side less than two right angles, then the two straight lines, if extended indefinitely, meet on that side on which the angles are less than the two right angles.

3 How to rephrase the fifth axiom

It is noticeable that Euclid's fifth axiom is more complicated than the other four, and mathematicians subsequently tried to rephrase it in order to simplify it. The most commonly known way of writing the fifth axiom is called Playfair's axiom, after the Scottish mathematician John Playfair, and says 'exactly one straight line can be drawn parallel to another through an external point.' This way of stating the fifth axiom in terms of parallel lines is why the fifth axiom is usually called the parallel postulate.

4 Consequences of the parallel postulate

The property of being able to draw unique parallel lines is used in a surprising number of other theorems in mathematics. This means that, if the parallel postulate were found to be false, all of these other results would be false, too. The most important of these is the result that the angles in any triangle sum to 180°. This is proved as follows: draw a triangle ABC with angles α, β and γ, and draw a line parallel to AB through C. Because the two lines are parallel, angle PCA = α, and angle QCB = β, and now we see that α, β and γ are on a straight line so must sum to 180°. Pythagoras' theorem relies on this result about triangles, and so it also depends on the parallel postulate.

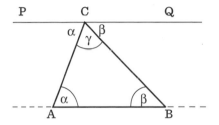

5 Whether parallel lines can be constructed

Euclid's definition of parallel lines was a pair of lines that never met. Mathematicians wondered if parallel lines could be constructed using an algorithm rather than just assuming they existed. For example, could a line parallel to another be constructed by joining together all points at a fixed distance from it and on the same side? The problem with this was that they could not prove that a line constructed this way would be straight. And indeed, the reason they could not find such a proof was that such lines are not always straight.

6 The example of spherical geometry
Although we all happily use Euclidean geometry without questioning it, we happen to live on a shape that does not obey Euclid's axioms: namely, a sphere. Here, the parallel postulate is definitely false. Straight lines on a sphere are great circles – circles whose centre is the centre of the sphere, such as the equator. The shortest distance between two points on a sphere is always a segment of a great circle, and this is why flight paths on a map always look curved. But we cannot draw two great circles parallel to each other: every pair of great circles crosses at two points. So the parallel postulate is false, and the method of constructing parallel lines at a fixed distance fails because it creates lines that are not great circles and, therefore, not straight.

7 Why spherical geometry was not thought a counter-example
Spherical geometry is clearly an example of the fifth axiom being false, so why did it not stop mathematicians trying to prove the parallel postulate? The problem was that it did not satisfy all of the other four axioms; namely, circles on a sphere cannot be constructed with any diameter, and straight lines on a sphere cannot be drawn infinitely long. The question was still whether a type of geometry could exist that did satisfy all four original axioms but in which the parallel postulate was false. As with many results in science, the answer came to three different people independently after 2000 years of being unsolved.

8 The three types of geometry
It turns out there are exactly three types of geometry: flat (Euclidean), spherical and hyperbolic. Hyperbolic geometry was first published by the Russian Nikolai Ivanovitch Lobachevsky in 1830, discovered independently by the Hungarian János Bolyai in 1832 (son of Farkas Bolyai, who discouraged him from studying the subject), and claimed to have been known by the great German mathematician Johann Carl Friedrich Gauss. In spherical geometry the diameter of space decreases as one moves away from the equator, causing straight lines to converge. In flat geometry distances are the same everywhere, so parallel lines are always the same distance apart. In hyperbolic geometry the diameter of space increases away from the centre, pushing parallel lines apart. Consequently, spherical geometry has no parallel lines, flat geometry has unique parallel lines, and hyperbolic geometry has infinitely many lines through a point parallel to a given one. Angles in triangles sum to more than 180°, exactly 180° and less than 180° in spherical, Euclidean and hyperbolic space, respectively.

9 How to draw hyperbolic geometry
The standard way that mathematicians draw hyperbolic geometry is using a model called the Poincaré disc. An example is the drawing shown below. Here, all the triangles are the same size, but they appear to get smaller as one moves out from the centre because distances are greater the further out we move. Straight lines are curves that meet the edge of the circle at right angles. Thinking of the circle as a field that is muddier the closer we get to the edge, we see that when moving from one point to another it is faster to take a path curving inwards to the less muddy region than to take an apparently straight line through the mud.

10 How crochet can model hyperbolic geometry
The Poincaré disc model was the standard way to draw hyperbolic geometry for over 100 years, until a Latvian mathematician called Daina Taimina realized that she could construct hyperbolic geometry very easily using crochet. Hyperbolic distances get bigger the further out you go, which she incorporated by adding in stitches at regular intervals as she crocheted. The resulting models looked very organic – like corals, fungi and leaves. This is explained by the fact that hyperbolic geometry creates surfaces that maximize surface area while minimizing volume, creating the perfect shapes for plants wanting to get the most nutrients from the sea or the most sunlight for photosynthesizing.

TALK LIKE A GENIUS

❝ Abraham Lincoln used to keep a copy of Euclid's *The Elements* in his saddlebag, saying "You cannot be a lawyer unless you understand what demonstrate means". ❞

❝ It is possible to draw a triangle on the Earth with three right angles. Start at the North pole, travel down to the equator, turn 90 degrees, go a quarter of the way around the equator, then turn again through 90 degrees and go back to the North pole. Each side of this triangle is a straight line because it is the shortest distance between points. ❞

❝ Euclidean space is the only geometry in which the ratio of the circumference of a circle to its diameter is a constant number, π. In spherical geometry the ratio is less than 3.14 and in hyperbolic geometry the ratio is greater than 3.14, with the exact value changing depending on how big the circle is. ❞

❝ The artist M.C. Escher created a series of four woodcuts called the *Circle Limit* figures, which were directly inspired by hyperbolic geometry. While corresponding with the mathematician Coxeter, Escher saw a picture of the Poincaré disc model and realized it showed how to tessellate a circle. The resulting images feature interlocking birds, fish and lizards, and – the most famous – angels and demons in an interwoven dance to infinity. ❞

WERE YOU A GENIUS?

1 FALSE – In flat geometry there is exactly one line parallel to a given one, going through a particular point. This is the parallel postulate.

2 TRUE – Triangles drawn on spheres have an angle sum greater than 180°, with larger triangles having a larger angle sum.

3 FALSE – Two lines are called parallel if they never meet. The two definitions coincide in Euclidean geometry, but not in spherical or hyperbolic geometry.

4 TRUE – On a sphere, the only straight lines are great circles, and two great circles always meet.

5 TRUE – Shapes with hyperbolic geometry have a larger surface-area-to-volume ratio than shapes with spherical or flat geometry.

THE BLUFFER'S SUMMARY

There are not one but three types of geometry: flat (Euclidean), spherical and hyperbolic, differing by how parallel lines behave.

Projective geometry

'Perspective is to painting what the bridle is to the horse, the rudder to a ship.'

LEONARDO DA VINCI

Filippo Brunelleschi, one of the founding fathers of the Renaissance, was the first to pioneer the use of perspective in drawing. Other people had noticed the simple fact that objects that are further away seem smaller, but Brunelleschi realized more precisely that lines that are parallel meet on the horizon. This idea of creating a vanishing point, or 'point at infinity', is the basis of projective geometry, which has become an extremely important area of mathematical study. Projective geometry lies at the heart of the Hodge conjecture, the solution to which is worth one million dollars, and is also invaluable in computer graphics and quantum physics.

What links Michelangelo with Tomb Raider? The answer lies in the maths of projective geometry.

ARE YOU A GENIUS

1 In Euclidean geometry, every pair of straight lines meets at a point.

TRUE / FALSE

2 When making a perspective drawing, an angle between lines in the drawing is the same as the angle between those lines in real life.

TRUE / FALSE

3 To describe a collection of parallel lines, the only information you need is the angle they make with the horizontal axis.

TRUE / FALSE

4 Every different collection of parallel lines meets the horizon at a different point in a perspective drawing.

TRUE / FALSE

5 It is possible to draw seven lines and seven points so that every point lies on three lines and every line contains three points.

TRUE / FALSE

TEN THINGS A GENIUS KNOWS

1 Why we add a point at infinity
On the Euclidean plane, any two points determine a line, but not every two lines determine a point. Most pairs of straight lines will cross at a single point, but some lines do not cross at all: namely, parallel lines. Taking inspiration from Renaissance artists, who allowed parallel lines to meet at a 'vanishing point' on the horizon, mathematicians realized they could add in points 'at infinity' that made the duality between lines and points work out perfectly. Euclidean geometry that uses these points at infinity is called projective geometry and is a model for how a 3-D scene looks when projected onto a 2-D screen.

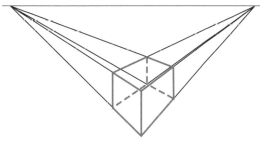

Vanishing point   Vanishing point

2 How projective geometry is different from Euclidean geometry
In a perspective drawing, angles and distances are not true to life. The square sides of a cube may look like trapeziums with the furthest side being shorter than the near side, even though, in reality, they are the same length. But some things are preserved under a projection; for example, straight lines. Points that lie on the same line in real life will lie on that same line in the perspective drawing. Similarly, in projective geometry there is no notion of distance or angles and it is the configuration of lines and points that is the most important property.

3 How to design coordinates for projective space
Mathematicians use a special set of coordinates to incorporate the additional points at infinity. These coordinates include all the original (x,y) coordinates of the plane, plus an extra point 'at infinity' for each set of parallel lines. Parallel lines are defined by the direction they face. So one set of parallel lines might all be at 45° in the plane, while a different set might be at 36°. Out of all the parallel lines at a 45° angle, exactly one of them goes through the origin (0,0). The coordinate system for projective space therefore needs to include all original points (x,y) plus a single representative of each line through the origin. A line through the origin is determined by a single point on the line, and any multiple of this point will determine the same line.

4 What homogeneous coordinates are
Coordinates for projective space are called homogeneous coordinates, and they use a third coordinate to incorporate the points at infinity. A homogeneous coordinate looks like $[x:y:z]$ with the restrictions that x, y and z cannot all be zero, and any multiple $[ax:ay:az]$ defines the same point. The standard (x,y) coordinates from the Euclidean plane are incorporated as $[x:y:1]$, while the point at infinity corresponding to the line going through (0,0) and (x,y) becomes $[x:y:0]$. So $[1:1:0]$ is the point at infinity corresponding to 45° parallel lines. Because of the rule that multiples of a point represent the same point, $[3:3:0]$ is exactly the same point in projective space as $[1:1:0]$. This corresponds to our intuition that the line joining (3,3) to (0,0) is the same line that joins (1,1) to (0,0).

5 A different way of seeing projective space
Homogeneous coordinates give us a new way of thinking about projective space. Each homogenous coordinate corresponds to a line through the origin in 3-D space: the rule that multiples of a coordinate define the same point is another way of saying that if one point is a multiple of another, then both define the same line through the origin. The 0 coordinate is excluded because this does not define a line at all. This way of thinking about projective space makes it easy to generalize to higher dimensions:

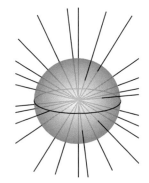

n-dimensional projective space is the space of lines through the origin in (*n* + 1)-dimensional Euclidean space.

6 The topological view of the projective plane

What does the projective plane look like as a shape? Thinking of the projective plane as lines in three dimensions gives us a clue here. If we imagine a sphere whose centre is the origin, then each line through the origin will intersect the sphere at antipodal (opposite) points. That is, each point in projective space is represented by a pair of antipodal points on a sphere. As a topological manifold (see page 90), the projective plane is created by gluing together each of these pairs of opposite points, creating a shape that twists through itself and is non-orientable. (See below – a projective plane sliced open.)

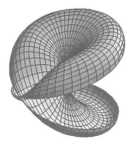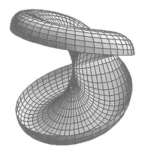

7 How conic sections are all the same

Conic sections are shapes that can be made by slicing through a pair of cones joined at their tips: circles, ellipses, parabolas and hyperbolas. They have been important since ancient times, explaining the motion of planets and comets, the trajectories of flying cannonballs and the shapes of bridges, towers and lenses. The conic sections all look very different: an ellipse is a closed shape, while parabolas and hyperbolas go off to infinity. But representing the equations for these shapes in homogeneous, or projective, coordinates shows that they are all really the same, except for where they cross the line at infinity. Ellipses do not touch the line, parabolas lie tangent to it (touching it once) and hyperbolas cross the line (touching it twice). So a hyperbola is really just an ellipse that has been to infinity and come back on the other side.

8 What the Fano plane is

Projective geometries can be concocted that only have a finite number of points and lines. The Fano plane is an example of this, containing only seven points and seven lines. Each line contains exactly three points and, by duality, each point lies on exactly three lines. A good puzzle is to try drawing such a configuration, bearing in mind that some of the lines may not look 'straight'! Finite projective spaces always have the same number of lines and points, but the question is open about which numbers can arise as the number of points in a projective geometry.

9 How projective geometry relates to group theory

The defining feature of projective geometry is that points that lie on a line will always lie along that line, even after any kind of transformation of the space. If we draw a finite projective geometry like the Fano plane, we can ask whether it is possible to permute the points around so that collinear points (that is, points on the same line) are still collinear. If it is possible, we can then ask how many ways there are of doing it. For the Fano plane, it turns out that there are 168 different permutations with this property, and these form an intricate mathematical object called the projective linear group.

10 Why projective geometry is needed in computer graphics

When playing a 3-D computer game, such as Tomb Raider, you do not experience a true 3-D world but only a picture on a 2-D screen. As you navigate around the virtual world, the image you see must change smoothly according to the direction in which you move the camera, whether that is panning across, turning or zooming in. Calculations for these coordinate changes are done using matrix multiplication, but not all of these effects can be represented by matrices if Cartesian coordinates are being used. However, the homogeneous coordinates of projective space are a perfect fit for doing easy translations, rotations and zooms.

TALK LIKE A GENIUS

❝ Artists have used clever types of projections to create optical illusions or hidden figures in paintings. The technique is called anamorphosis, and one of the most famous examples is in a painting called *The Ambassadors* by Hans Holbein the Younger. Viewed from the front, the scene is a calm one of two men standing by a collection of scientific instruments, but viewed from a point on the bottom left, a skull suddenly appears! ❞

❝ Projective geometry is used on road markings to make signs easier to read for a moving driver. Painted warnings such as 'SLOW' or 'KEEP CLEAR' are drawn in an elongated fashion so that they appear in the correct aspect ratio for a driver seeing them from a distance. ❞

❝ In quantum theory, the wave function of a particle is represented as a vector in something called a Hilbert space. Multiplying the wave function by a number does not change the physical state that the particle is in. This means that wave functions live naturally in projective space, where taking multiples of a coordinate does not change the point it represents. ❞

THE BLUFFER'S SUMMARY

Projective geometry models the geometry in perspective drawings, where parallel lines meet 'at infinity' and where distances and angles may be distorted from their real-life values.

The shape of our universe

'Do not worry about your difficulties in mathematics. I can assure you that mine are still greater.'

ALBERT EINSTEIN

Despite the myth that Einstein was bad at mathematics, he was actually a brilliant mathematician whose papers were filled with new and complex geometrical ideas. In order to formulate his theory of general relativity he needed a new branch of mathematics – Riemannian geometry – which allowed him to describe how the universe curved in the presence of matter. Riemann's work had revolutionized geometry in the 19th century, but little did he know that in just 60 years it would revolutionize our whole understanding of physics.

How can we begin to imagine the shape of our four-dimensional universe? The answer starts by understanding the curvature of the Earth.

ARE YOU A GENIUS

1 The area of a circle is πr^2 no matter what space the circle is drawn on.

TRUE / FALSE

2 To find the shortest flight path from London to New York, pilots draw a straight line between the cities on a map.

TRUE / FALSE

3 If the Earth were a cylinder instead of a sphere, we could draw accurate maps with no distortions.

TRUE / FALSE

4 A stick carried on a path around the Earth, carried at the same angle throughout the journey, will return parallel to how it began.

TRUE / FALSE

5 The curvature of our universe will determine its future.

TRUE / FALSE

TEN THINGS A GENIUS KNOWS

① What Riemannian geometry studies
Just as Euclidean geometry is the study of flat space, Riemannian geometry is the study of curved space. When finding the distance between London and New York, how do we know what the shortest path between them is? How do we measure the area of a country, taking into account that it lies on a sphere? We learn in school that the area of a circle of radius r is πr^2 but this is only true in flat space: circles drawn on a sphere have a far more complicated formula that depends not only on the radius of the circle, but also on the radius of the sphere. Riemannian geometry answers these questions, and goes further to examine spaces whose curvature may vary, and how curvature works in higher dimensions.

② How to measure curvature
Curvature is easiest to understand in one dimension: that of a line. We can measure how curved a line is at a particular point P by looking at the largest possible circle that approximates the curve at that point. The smaller the circle, the more curved the line is. If the circle has radius r, the curvature at P is defined to be $1/r$. Curvature is also marked as being either positive or negative. If, as we travel along the curve, the line joining P to the centre of the circle is moving anticlockwise, then the curvature is positive, but if it is moving clockwise then the curvature is negative.

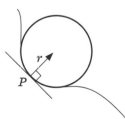

③ What Gaussian curvature is
Gaussian curvature is a way of measuring the 2-D curvature of a surface, based on our understanding of curvature in one dimension. At a particular point, we imagine cutting the surface with a plane and looking at the cross-sectional curves that result. The various cross-sections will all have different curvatures, and the product of the maximum and the minimum is called the Gaussian curvature. This measurement of curvature is intrinsic in that it can be measured by a creature living in the surface. It does not depend on the way that the surface is embedded in the surrounding space.

④ The three types of curvature
For curves there can be positive, negative and zero curvature, and the same is true of surfaces. When all the cross-sections at a given point curve in the same direction as each other, as on a sphere, the curvature will be positive because we are multiplying positive by positive or negative by negative. If one cross-section is curving up while the other is curving down, as in the saddle shape pictured, the resulting curvature will be negative. And if one of the cross-sections is flat, as on a cylinder, then the curvature will be zero. These three types of geometry are called spherical, hyperbolic and flat (or Euclidean), respectively.

⑤ Why curvature in three and four dimensions is important
John Wheeler summarized Einstein's theory of general relativity as 'Spacetime tells matter how to move; matter tells spacetime how to curve'. The presence of matter creates distortions in spacetime, whose geometry then influences the paths that moving objects take, and this is what we experience as gravity. For Einstein to create his theory, he needed to know how to measure curvature in not only the three dimensions of space, but the four dimensions of spacetime. The kind of curvature needed for the study of relativity was more complicated than could be captured by a single number, as with Gaussian curvature, so a new approach was developed based on the concept of parallel transport and connections.

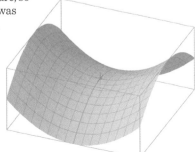

6 **How parallel transport defines curvature**
Imagine a relay race in which a runner holds a straight baton that she wishes to give to another runner on the other side of the track. If the track is completely flat, and if the runner holds the baton so that it is always parallel to its starting angle, then it does not matter which path she takes on the race – the baton will arrive being held at the same angle. Contrast this with a race in which the baton must be delivered to the Earth's North pole: here different paths will result in the baton being held at different angles. A runner going directly from A to N in the picture will give a different result from that of first taking the baton from A to B, and then from B to N. The difference is because of the curvature of the Earth, and this difference is actually a measure of the curvature.

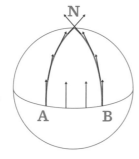

7 **The fundamental theorem of Riemannian geometry**
A connection on a manifold is a set of instructions for transporting data along curves. An important example of such data is a tangent vector. A connection defines a collection of special curves, called geodesics, that are curves along which tangent vectors remain tangent as they are transported. But geodesics have another definition based on the measurement of distance on the manifold: a geodesic is the shortest path between two points. The fundamental theorem of Riemannian geometry states that there is exactly one connection on a manifold so that the two definitions of geodesic curves are the same. This connection, called the Levi-Civita connection, is used to define the Riemann curvature tensor – the central tool in the theory of general relativity.

8 **The Gauss-Bonnet theorem**
One of the most striking and important theorems about curvature of surfaces is the Gauss-Bonnet theorem, which relates the geometry of a shape to its topology. It says that if we add up the Gaussian curvature at every point on a closed surface with g holes in it, then the answer will be 2π multiplied by $(2 - 2g)$. So the curvature entirely determines the global shape of the surface. If we add up the curvature and find that it is zero, for example, the Gauss-Bonnet theorem tells us that the surface must have one hole in it – a torus, or doughnut. A similar result is true in higher dimensions, relating the number of holes in a manifold to the total curvature as measured by the Riemann curvature tensor. This means that if we could somehow determine the curvature everywhere in the universe, we would know something about its shape.

9 **What the curvature of the universe is**
The total curvature of the universe, which relativity tells us is governed by the quantity of matter in it, is quantified by the density parameter Ω. This density parameter is the average density of matter in the universe divided by the amount of mass energy needed for the universe to be flat. So if $\Omega = 1$, then the universe is flat, if $\Omega > 1$ then the universe is positively curved, and if $\Omega < 1$ then the universe is negatively curved. We can try to find out which geometry we are in by either trying to calculate the total amount of mass in the universe, or by trying to measure the angles inside geodesic triangles. So far, measurements of both types indicate that the universe is very close to being flat.

10 **Implications for the future of the universe**
Finding out what type of total curvature the universe has will tell us about its shape and its future. A positively curved universe must be finite in extent. Gravity would eventually stop the expansion of such a universe, and collapse it back to a point in a 'big crunch'. A negatively curved universe would expand forever at an accelerating rate, causing a 'big freeze'. A flat universe would expand forever, but at a decelerating rate, and could be either finite or infinite depending on the number of holes. Astronomers and mathematicians study the cosmic microwave background radiation for patterns that may reveal which scenario we are in.

TALK LIKE A GENIUS

❛ Riemannian geometry is important for pizza lovers. A slice of pizza has a total Gaussian curvature of zero. When picking up a slice, the pointy end tends to flop downwards, making it hard to eat. This can be counteracted by bending the slice upwards along the crust, which forces the pointy end to become straight again, in order to maintain the total zero curvature. ❜

❛ Gaussian curvature explains why we can never draw a flat map of the Earth without distorting distances. A flat map has zero curvature, while the Earth has positive curvature, so no piece of paper can wrap around the sphere without crumpling or stretching. ❜

❛ Scientists believe that there is no "edge" to the universe, but this does not restrict whether the universe could be finite or infinite. For example, the surface of the Earth is a shape that is finite but has no edge. A flat universe could be finite if it had the shape of a torus, although, if this were the case, we would expect to see repetitive patterns in the cosmic microwave background radiation, as light travelling in one direction would reappear from another direction. So far no such patterns have been spotted. ❜

WERE YOU A GENIUS?

1 FALSE – On curved spaces the formula for the area of a circle varies, depending on the underlying curvature of the space itself.

2 FALSE – Most maps of Earth distort distances in trying to project the curved Earth onto flat paper, so straight lines on the map are no longer shortest distances (though a few special projections preserve this feature).

3 TRUE – The cylinder is a shape with zero Gaussian curvature, as is a sheet of paper, so it could be accurately drawn on paper without distortion.

4 FALSE – The curvature of the Earth causes the stick to change direction relative to its starting angle.

5 TRUE – The universe may expand forever or collapse back to a single point, depending on its total curvature. This, in turn, depends on the amount of mass and energy in it.

THE BLUFFER'S SUMMARY

Riemannian geometry studies curved spaces. It is a key component in the formulation of general relativity and the study of the shape of our universe.

Wallpaper patterns and Penrose tilings

> 'I am driven by the irresistible pleasure I feel in repeating the same figures over and over.'

M.C. ESCHER

The talented artist M.C. Escher is well known for creating beautiful and intricate tiling patterns: horses, lizards, angels and demons that fit together effortlessly as they travel out to infinity. Yet, despite the seemingly endless variety of tiling shapes, mathematicians have proven that there are really only 17 possible types of pattern for your wallpaper – that is, unless you want a unique wallpaper that never repeats itself, in which case mathematicians have a solution for that, too. And while wallpaper patterns may seem relevant only to interior designers, these genius ideas anticipated a scientific discovery that won a Nobel Prize for Chemistry.

The symmetries of wallpaper patterns and pavement tilings have been the inspiration for genius breakthroughs in material science.

1 A given collection of tiling shapes can only create one type of tiling pattern.
TRUE / FALSE

2 It is impossible to create a tiling pattern without symmetry.
TRUE / FALSE

3 Regular pentagons, by themselves, cannot create a tiling pattern.
TRUE / FALSE

4 Every tiling pattern can be translated (that is, shifted), so that the tiles once again line up exactly as before.
TRUE / FALSE

5 Crystals have molecular structures that are tiling patterns in three dimensions.
TRUE / FALSE

TEN THINGS A GENIUS KNOWS

① What a tiling pattern is

A tiling pattern is a way of using shapes (of a fixed number of different types) to cover a flat surface, or plane, in such a way that there are no gaps or overlaps. We are used to seeing such patterns on pavements, on wallpaper and on our bathroom walls. When mathematicians think about tiling patterns, they imagine the tiles extending out to infinity in every direction. A symmetry of such a pattern is any transformation that leaves the pattern looking exactly as it did before. In the Roman mosaic shown, the pattern looks the same when reflected through either of the dashed lines, or by rotations of 60° through the point at which the dashed lines meet.

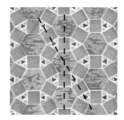

② The difference between periodic and aperiodic tilings

A periodic tiling is one with translational symmetry. That is, one could trace the shape of the tiling onto paper, and then shift the paper to make the drawing once again line up exactly with the original. An aperiodic tiling does not have this property: no shifting of the paper, in any direction and for any distance, will cause the drawing to line up again with the tiles. Some tiles can create both periodic and aperiodic tilings, but a set of tiles is called aperiodic if they can *only* create aperiodic tilings. Initially considered impossible, different examples of such tilings were discovered in the 1960s and 1970s, most famously by Sir Roger Penrose.

③ When two tilings are the same

Mathematicians consider two tiling patterns 'the same' if they have the same collection of symmetries. The Persian pattern on the right looks very different from the Roman mosaic above, but both have the same collection of symmetries: translations, reflections and rotations. Every action that could be done to one picture

leaving it looking the same could also be done to the other picture. The Egyptian ceiling (right) however, has a different kind of tiling. Here we see rotational symmetry of order four, and there are no possible lines of reflection. The collection of symmetries that define a 2-D tiling pattern are called its 'wallpaper group'.

④ How many wallpaper groups there are

Mathematicians have proved that there are only 17 different wallpaper groups – that is, only 17 different symmetry combinations that define a tiling pattern. The standard notation for a wallpaper group is called the Hermann-Mauguin notation and uses four symbols. The first is either a 'p' or a 'c', depending on whether the main axis of rotation/reflection runs along the sides of the main tile or through its centre. The second symbol is a number that gives the highest order of rotational symmetry. The last two symbols indicate the presence of a mirror (m) or glide (g) (a reflection combined with a translation) along the two main axes of the tiling. Using this notation, the 17 groups are p111, p1m1, p1g1, c1m1, p211, p2mm, p2mg, p2gg, c2mm, p311, p3m1, p31m, p411, p4mm, p4gm, p611 and p6mm.

⑤ Which rotations are allowed in a tiling

In the 17 wallpaper groups, the only rotation orders allowed are one, two, three, four and six. You will never see a wallpaper pattern in someone's home that has a five-fold symmetry. The reason for this is that if a five-fold symmetry existed, it would imply that the basic tile used to make the pattern would be infinitely small. However, it turns out that if we remove our insistence that there is translation symmetry in the pattern, then five-fold symmetries *can* exist after all. This was beautifully demonstrated in Roger Penrose's aperiodic tiling, using two shapes known as 'kites' and 'darts', and would later rock the foundations of crystallography.

⑥ How kites and darts create an aperiodic tiling

A Penrose tiling is made up of kites (below left) and darts (below right) together with rules on how the shapes are allowed to fit together. The shapes naturally fit together to make a rhombus, or diamond, but this is prohibited in order to prevent a periodic tiling. In the picture shown, the kites and darts are only allowed to fit together so that the dots are next to each other. There are seven different ways these two tiles can fit together around a point. In particular, there are two ways to start a Penrose tiling that creates a pattern with five-fold rotational symmetry. This symmetry only occurs at one point in the (infinite) Penrose tiling.

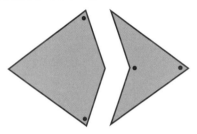

7 How and why the golden ratio appears in a Penrose tiling

The golden ratio appears everywhere in a Penrose tiling. It is the ratio of the long to short sides of both the kites and darts; the ratio of the area of the kite to the area of the dart; and the ratio of the number of kites to the number of darts in any tiling. What is it that makes this magic number appear so often? The answer lies in the way the kites and darts are generated: as regions of a pentagon inside another pentagon. The golden ratio is the ratio of the length of a chord (diagonal) of a regular pentagon to the length of a side, which is why it appears in the side lengths of the kites and darts.

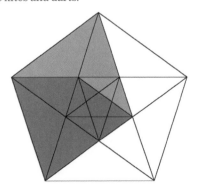

8 The problem of identifying a Penrose tiling

Using kites and darts, there are infinitely many ways to arrange them to create a tiling. If a friend shows you a drawing of part of a tiling, is it possible to identify which tiling it is? The answer is no: any finite region occurs in every possible Penrose tiling, and it occurs infinitely often in each one. Regions with identical tiles are surprisingly close together. If you have managed to locate your friend's drawing in a Penrose tiling, you will find another copy of that region within a distance of twice the diameter of the drawing. This seems incredible for a pattern that has no translational symmetry.

9 How a Penrose tiling is self-similar

It is possible to subdivide the kites and darts into scaled-down copies of kites and darts (using some sneaky half-darts, too). If we continue this process, which is called deflation, we create ever more intricate Penrose tilings that have a fractal self-similarity. Applying this process in reverse, which is called inflation, allows us to tile ever larger areas with Penrose tiles, and was part of the original proof that kites and darts can create an infinite tiling. Many different families of aperiodic tilings have been created by using this method of substitution.

10 How the maths of tiling relates to crystals

A crystal is made up of repeating patterns of molecules, so it is a 3-D version of a wallpaper pattern. In 1892, Russian crystallographer Evgraf Fedorov and German mathematician Arthur Moritz Schoenflies created a proof that there are exactly 230 different 3-D wallpaper groups (space groups). Each of these 230 symmetry patterns has since been shown to exist in real crystals. In the 1980s, a chemist named Dan Shechtman made the bizarre discovery of a crystal that contained a five-fold symmetry, which Fedorov and Schoenflies had shown to be impossible. It took decades for scientists to accept that crystal structures could exist without translational symmetry. Shechtman won the Nobel Prize for Chemistry in 2011, and his discovery has been named a quasicrystal. Quasicrystals occur both naturally and synthetically, and have the potential for making materials such as bulletproof armour, non-stick coatings for frying pans and invisibility cloaks.

❝ The result that there are 230 space groups was actually found before the proof that there are 17 wallpaper groups. Although many cultures around the world had found the 17 patterns, the crystallographer Fedorov realized that nobody had proved that no other patterns were possible. So after he solved the 3-D case he went back and proved the (much easier) 2-D case as well. ❞

❝ It has been claimed that all 17 wallpaper groups can be found at the Alhambra Palace in Spain, though other people refute this on the grounds that researchers have not been consistent in whether or not they include colour as part of their classification. ❞

❝ Regular pentagons cannot make a tiling pattern, but an irregular pentagon can. In 1985, 14 such tiling pentagons were known – some discovered by amateur enthusiasts. A 15th pentagon was discovered in 2015, and in 2017 Michaël Rao submitted a proof that there were none left to find. But his proof only focused on convex pentagons – there are potentially still non-convex pentagonal tiles to be discovered. ❞

❝ An obvious question after the discovery of Penrose tiles was whether there exists a single tile that is aperiodic. Such a tile was named an 'einstein', as in German this means 'one stone'. It is unclear whether an einstein has been found, as mathematicians are still arguing about the properties that such a tile would need to have in order to qualify. ❞

WERE YOU A GENIUS?

1 FALSE – A set of tiles can (usually) create multiple tiling patterns, each with a different set of symmetries.

2 FALSE – There are aperiodic tilings without translational, rotational or reflectional symmetry.

3 TRUE – The angles in a pentagon mean that they cannot fit together to tessellate the plane, though pentagons can combine with other shapes to create tiling patterns.

4 FALSE – Aperiodic tilings exist without any translational symmetry; Penrose tilings are one example of this.

5 TRUE – There are 230 different symmetry types for crystals, and nature has found all of them.

THE BLUFFER'S SUMMARY

There are only 17 ways to tile your bathroom floor with a repetitive pattern. There are also additional aperiodic tilings that will never repeat, even if your floor were infinitely big.

The sphere-packing problem

'Nature uses as little as possible of anything.'

JOHANNES KEPLER

What is the most efficient way to stack cannonballs or oranges? A question that soldiers and greengrocers solved without thinking about it turned out to be surprisingly difficult for mathematicians. Although the problem was finally considered solved in 2017, it has left behind it a trail of unanswered questions about arranging shapes in space, including how to find the most efficient bubble structures, whether spheres might be the worst possible shapes, and why dimension eight is a Goldilocks dimension for sphere packing.

Sometimes it takes a genius to figure out how to pack a suitcase – the same genius that's needed to design nano-materials and folding telescope lenses.

1 Given a box of particular dimensions, it can be more efficient to pack spheres in it randomly rather than in a regular pattern.
TRUE / FALSE

2 The greengrocers' method of stacking oranges is the most efficient way of packing spheres in space.
TRUE / FALSE

3 It is possible to get eight spheres of a given size to touch another sphere of the same size, but no more.
TRUE / FALSE

4 The reason bees create honeycombs using hexagons is because it is the shape that requires the least wax to create the cells.
TRUE / FALSE

5 The circle is the most inefficient shape for packing, in terms of the empty space still left over.
TRUE / FALSE

TEN THINGS A GENIUS KNOWS

1 What the Kepler conjecture is
In 1611, the German astronomer Johannes Kepler considered the best way of stacking cannonballs on a ship. More generally, he wondered how best to arrange identically sized spheres in order to fill out space. In two dimensions, the corresponding problem would be how best to arrange circles of the same diameter. The 'obvious' solution to the 2-D problem is to arrange the circles in offset rows so that the centres of one row are halfway between the circles on the row beneath. The 3-D solution, which became known as Kepler's conjecture, was to arrange a layer of spheres as in the 2-D solution, and then stack subsequent layers offset, so that each sphere would lie in the indentation made by three spheres beneath it.

2 The difference between regular and irregular packings
The difficulty in proving the Kepler conjecture lay in the possibility that the most efficient packing could be achieved by arranging the spheres in an irregular fashion. Gauss had proved in 1831 that Kepler's 'greengrocer' packing was the most efficient arrangement if the centres of the spheres had to form a regular lattice. But it was known that when

circles were packed into a finite area, irregular packings were sometimes better. In the picture shown, 15 circles fit into the box. Was it possible that such an irregular solution might exist for the infinite case as well?

3 How Thales solved the Kepler conjecture
The American mathematician Thomas Thales began work on the Kepler conjecture in 1992, building on work by another mathematician, Laszlo Fejes Toth, who had shown that it was possible to solve the Kepler conjecture by checking a finite number of cases. Over the following six years, Thales designed a computer program to check these cases, but the code was so complicated that, even in 2003,

mathematicians still had not been able to verify that it was 100 per cent correct. Over the following ten years Thales went on to produce a more formal proof of the conjecture, which has been checked by automated theorem-proving assistants and was accepted by the mathematical community in 2017.

6 Ways in which 8-D space behaves oddly
Imagine arranging four equally sized circles in a square formation. There will be a small hole in the centre into which we could fit another circle, but this new circle would be far smaller than those we started with. Similarly, we can arrange eight equally sized spheres in a cube, and there will be a hole in the middle that can fit in another sphere, but only a smaller one. We can continue this pattern in higher dimensions, but something very strange happens in dimension eight. When we arrange spheres in this dimension, the hole in between the spheres is large enough that we can fit in a new sphere that is the same size as the ones we started with, leading to a way of doing sphere packing that is far more efficient than in other dimensions. In 2016, Maryna Viazovska found the best sphere packing in dimension eight (and also 24) using these ideas, while the best sphere packings in all other dimensions are still unknown.

5 How circle packing relates to origami
Circle packing has a surprising use in origami design, most famously demonstrated by physicist and artist Robert J Lang. In the early 1990s, Lang developed a computer program that could design a folding pattern for any origami figure, no matter how intricate. The program starts with a tree diagram that provides the underlying skeleton of the structure; for example, an origami human would need a branch for the main torso, a branch for each limb and one for the head. A more realistic human would have additional branches for hands and feet, or even fingers and toes! The program takes this tree and applies a circle packing algorithm to generate the crease patterns. Lang's designs are used in engineering as much as art, creating air bags and telescope lenses, which need to pack down small and then open up at exactly the right time and in the right way.

6 Kissing numbers

The Kepler conjecture is related to a problem about 'kissing numbers', which asks how many spheres (of equal size) can be arranged to touch a single sphere. In two dimensions we can arrange for six (non-overlapping) circles to simultaneously touch another circle, and in three dimensions we can get 12 spheres to touch a central sphere. (This was conjectured by Isaac Newton but not proven until 1953). The kissing-number problem influences how atoms can be arranged in a crystal structure. The answer in four dimensions was only found in 2003, and is unknown in higher dimensions except for eight and 24, just as in the sphere-packing case.

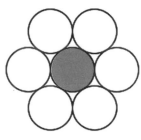

7 How to optimize bubbles

In discussing the sphere-packing problem, we are arranging shapes that will never fit perfectly together, and so there will always be gaps between them. A related question asks about how bubbles might arrange themselves into a foam in the most efficient way, giving a structure with no gaps at all. It was Lord Kelvin who asked the more precise question of how we could partition space into cells of equal volume, so that the surface area of touching cells would be as small as possible. In two dimensions the answer is the honeycomb (proved by Thales in 1999), and in three dimensions Kelvin conjectured that the cell shape would be that of a truncated octahedron, a symmetric shape whose faces consist of hexagons and squares.

8 What the Weaire-Phelan structure is

Mathematicians spent over 100 years trying to prove the Kelvin conjecture about the most efficient foam structure. It came as a great shock when, in 1993, physicist Denis Weaire and his student Robert Phelan discovered new bubble shapes that were even better at filling out space than Kelvin's truncated octahedron. They used two different kinds of cells that had the same volume as each other. One cell is called a pyritohedron and has faces that are irregular pentagons. The other is a 'truncated hexagonal trapezohedron', which has both hexagonal and pentagonal faces. Unsurprisingly, nature had already found this solution, which can be seen both in bubbles and in the structure of atoms. Mathematicians have still not decided whether this is the best possible solution.

9 How to pack spheres of different sizes

In the real world, a far more common problem about sphere packing is when the spheres to be packed come in different sizes. One example occurs in material science, where scientists look at how to add nanoparticles to a structure in order to strengthen it. How big should the particles be to achieve the best packing of molecules? If the diameter of one of the spheres is less than about 40 per cent of the first one, then the small spheres can fit into the gaps between the big spheres. But if they are larger, then whole new packing arrangements need to be discovered. Breakthroughs have been found for particular ratios of spheres, but in general the problem is very difficult even for two sizes of sphere, and almost impossible when the spheres come in more than two sizes.

10 Whether spheres are the worst possible packing shapes

Even with the best possible packing arrangement of spheres, more than one-quarter of the space will be left empty. 'Ulam's packing conjecture' claims that no other shape will ever be as inefficient as this at filling space. That is, if we take any other (convex) 3-D shape, it will always fill out more space when packed than spheres will. This has been shown to be true for shapes that have a certain kind of symmetry, but there are still an infinite variety of irregular shapes that could yet break the conjecture. In two dimensions the corresponding result is not true: regular octagons and heptagons are even worse at filling out space than circles are.

TALK LIKE A GENIUS

❝ The Weaire-Phelan structure, which is the most efficient foam structure currently known, was used as the inspiration for the design of the Beijing National Aquatics Centre, built for the 2008 Olympics. ❞

❝ In dimension 24 the sphere packing is so compact that each sphere is surrounded by 196,560 other spheres. This fact was used to develop an efficient error-correcting code, called the Golay code, which allowed scientists to communicate with the *Voyager* probes as they travelled around the solar system. ❞

❝ Kelvin's truncated octahedron has a very special property when it is embedded in four-dimensional space. The coordinates of its 24 vertices can be found by writing down all the possible permutations of (1, 2, 3, 4). This property means that the shape is the "permutohedron" of order four. The permutohedron of order three is the hexagon, which is the best two-dimensional tiling shape, though it is unlikely that Kelvin knew of this connection. ❞

WERE YOU A GENIUS?

1 TRUE – This can be true for some box and sphere sizes, but becomes false once we start thinking about infinitely sized boxes.

2 TRUE – This was only proved true in 2014, although it was conjectured by Kepler in 1611.

3 FALSE – It is possible for 12 spheres to touch another sphere of the same size.

4 TRUE – The best way to partition space into equal-sized cells using the least material is by creating hexagons.

5 FALSE – Regular heptagons and octagons are even worse at filling space than circles are.

THE
BLUFFER'S
SUMMARY

Mathematicians took 400 years to prove that greengrocers were stacking oranges in the most efficient way, opening up questions about shape-packing that are relevant to nano-engineers, honeybees and bubbles.

Topology

'A topologist is a person who cannot tell the difference between a doughnut and a coffee cup.'

ANONYMOUS

Mathematicians try to simplify problems as much as they can, getting rid of any unnecessary information. In the 18th century, they realized that some problems in geometry don't require any of the standard information, such as measurements of size, angles, position or curvature. Instead, what was needed was a qualitative description of shapes – an understanding of their intrinsic structure. Hence, the subject of topology was born. It came into its own during the 20th century and has since become one of the major branches of mathematics.

Sometimes being a genius means knowing when to throw away information. Hence, geometry without angles or sizes becomes the new subject of topology.

ARE YOU A GENIUS
?

1 Topologically, a sphere is the same as a cube.

TRUE / FALSE

2 An infinity symbol is topologically the same as a circle.

TRUE / FALSE

3 There are different ways of measuring the distance between two points.

TRUE / FALSE

4 A shape that looks flat at each point cannot be topologically interesting.

TRUE / FALSE

5 The Earth, being a sphere, cannot be represented accurately on any flat map or collection of flat maps.

TRUE / FALSE

TEN THINGS A GENIUS KNOWS

1 The solution to the seven bridges of Königsberg

In 1736, Leonhard Euler invented a new area of mathematics by solving an ancient problem. The city of Königsberg (now Kaliningrad) was formed of four pieces of land joined together by seven bridges. The question was whether it was possible to make a tour of the city, crossing each bridge once only. Euler's breakthrough was to realize that the exact positions of the rivers and the sizes of the islands was irrelevant: all that mattered was how many bridges connected each pair of land areas. To perform a tour, a person would need an even number of bridges adjoining each land area so that every time they entered the area they could also leave again. Königsberg did not have this property and so a tour was impossible. In solving the problem, Euler had used a kind of mathematics in which precise measurements of size or position were irrelevant, and continuous deformations didn't change the problem. This is topology.

2 Why topologists consider doughnuts and coffee cups to be the same

In topology, two shapes are considered 'the same' if one can be continuously deformed into the other without any breaking, cutting or gluing. A round ball of plasticine or clay can be moulded smoothly into a cube, so topologically the sphere and the cube are the same. But the same ball of plasticine cannot be made into a coffee cup because, in order to make the handle, one would need to join two separate bits of plasticine together, or else make a hole. However, if the plasticine started out in the shape of a coffee cup with one handle, it could be morphed into a doughnut or bagel, as the hole made by the handle would become the hole in the centre of the doughnut. Size, angles, corners and curvature are all irrelevant to a topologist: the important thing is to capture the intrinsic structure of an object.

3 How we capture the idea of continuity

Since topology is about the study of things changing continuously, mathematicians needed to get to the heart of what continuity was. The basic idea is that, if an object is changing continuously, then points that are close together will remain close together after a small time period. Cutting a piece of string is not a continuous motion because two points that are next to each other become far away after the cut. But the idea of 'close by' might also depend on the situation. For example, two towns might be close together as the crow flies, but the only way of getting from one to the other is on the train going via London. An earthquake may push the towns closer by one measure but move them far apart by the other measure. So different ways of measuring distance will change our notion of continuity, and will therefore change the topology.

4 What open sets are

To get around the difficulty of having to define a notion of distance before they could define 'closeness', mathematicians came up with the much more general notion of an 'open set'. Intuitively, an open set is a collection of points that are considered approximately the same. The more open sets we define, the greater accuracy we have in approximating points, because we can recognize two points as being distinct if they are in different open sets. Open sets satisfy a collection of technical rules, in order to ensure that they follow our instincts for how distance measurements should behave. For example, if A is close to B, and B is close to C, then by some measure A is also close to C.

5 What a topological space is
A topological space is a set of points together with a list of the open sets. The relationships between the open sets, and in particular how they overlap each other, is what determines the shape of an object. As an example, consider a length of string with four knots – A, B, C and D – tied at regular intervals. Points that lie between the same two knots are roughly similar, so we say that the open sets of the space are the three intervals A to B, B to C and C to D. Because the knots B and C are each in two intervals at once, we see that they are joining together the sections, forming a line. If there were an additional open set consisting of A and D, this would tell us that the two ends of the string were glued together and that the space would be a circle.

6 What a manifold is
While most topological spaces are exotic and strange, the ones of greatest interest to us are often the ones that look most like the space we are familiar with, and that is Euclidean (flat) space. In one dimension, Euclidean space is a straight line; in two dimensions it is a plane and in three dimensions it is like the regular space we live in. We can create topological examples by using Euclidean space as a basic building block, gluing together small pieces of Euclidean space to build interesting structures. These structures are called manifolds. At every point on a manifold, space looks Euclidean, even though globally it may be curved and twisted.

7 Some examples of manifolds
We live on a manifold. It took people a long time to realize that the Earth was a sphere, because in small regions it does seem to be flat. Another 2-D manifold is a torus, or doughnut. A circle is a 1-D manifold, because if you zoom in far enough on a segment of a circle it looks like a straight line. The infinity sign ∞ is not a manifold because, at the point where the curve intersects itself, it does not look like a line.

8 How to describe a manifold
We cannot draw a map of the world accurately on a single piece of paper, because the Earth is globally curved while the paper is not. But because the Earth is a manifold, we can make an atlas of different maps, which, together, cover the Earth.

For example, we could have an atlas for the Earth with two maps in it: one for the northern hemisphere and one for the southern hemisphere, with points on the equator represented in both. In a similar way, there is an atlas for every manifold: a collection of local flat maps, together with information on how the different maps overlap. It is the overlapping information that determines the global shape of the manifold.

9 What topological invariants are
The basic question in topology is to decide whether or not two spaces are homeomorphic – that is, having the same topological properties. One technique is to find a property of a topological space that does not change as the space is deformed. Such a property is called an invariant, and if two spaces have different values of the same invariant, then they must be different spaces. One invariant is the notion of connectedness. A space is path-connected if any two points can be joined by a path lying within the set – so the word PATH is not path-connected because the points on the P cannot get to the points on the A without going outside the set. Another invariant is the number of points that can be removed from a space before it stops being path-connected. So the letter D requires two points to be removed before it stops being path-connected, while the letter S requires only one, meaning they are not homeomorphic.

10 Where topology is used in science
The Nobel Prize in Physics in 2016 was awarded to three scientists who used topology to study unusual states of matter, such as superconductors and superfluids. When electrons moved in very thin layers, their conductance was found to change in whole-number steps, which turned out to be a manifestation of a topological invariant. In biology, scientists are using topology to study the connectivity of the brain, and how this changes as we experience new situations. Knowledge about the topology of the brain is also used in the analysis of MRI images to remove errors, such as the appearance of holes that should not be there.

TALK LIKE A GENIUS

❡ The word 'topology' was coined by Johann Benedict Listing in the 19th century and comes from Greek, meaning 'the study of place'. It is not to be confused with 'topography', which is the study of the Earth's shape and features, and means 'description of place'. ❡

❡ To demonstrate your topological skills and win a bet, challenge a friend to pick up two ends of a scarf and then tie it into a knot without letting go of the ends. If they pick up the ends normally they will create an unknot and nothing they do will be able to change that. To win the bet, fold your arms across your chest before picking up the scarf. ❡

❡ For a mathematically interesting breakfast, cut your bagel in half by twisting the knife through a full revolution as you move around the bagel. The result will be two interlinked bagel halves, creating twice as much surface area for spreading your cream cheese. ❡

❙ **TRUE** – A sphere made of plasticine can be smoothly moulded into a cube, so they are the same topologically.

❷ **FALSE** – The infinity symbol intersects itself, so is different from a circle.

❸ **TRUE** – Different ways of measuring distances will result in different notions of continuity, and thus different topologies.

❹ **FALSE** – Manifolds are topologically interesting shapes that are created by gluing together small pieces of flat (Euclidean) space.

❺ **FALSE** – As the Earth is a manifold, it can be represented by an atlas of flat maps; for example, one for each hemisphere.

THE BLUFFER'S SUMMARY

In topology – a kind of geometry in which squares are the same as circles and doughnuts are the same as coffee cups – it is the intrinsic structure of the shapes that matter, rather than size, angles or position.

Möbius strips and non-orientable surfaces

'Everyone knows what a curve is, until he has studied enough mathematics to become confused through the countless number of possible exceptions.'

FELIX KLEIN

Raise your right hand. Now your left. If you travelled to a different country, would you still raise your hands in the same order? What if you travelled around the universe and back? It might seem inconceivable that right and left should ever switch places, but 19th-century mathematicians discovered weird universes in which this could happen. Non-orientable shapes have remained a source of fascination ever since.

The twisted nature of the Möbius strip turns our ideas of right and left upside down – can you wrap your head around it?

1 A Möbius strip has only one side and one edge.
TRUE / FALSE

2 It is possible to find a shape that gets bigger when it is cut in half.
TRUE / FALSE

3 A circular strip of paper containing a full twist becomes two interlocking loops when it is cut in half.
TRUE / FALSE

4 Every closed surface (that is, one without edges) has both an inside and an outside.
TRUE / FALSE

5 If a person woke one day to find they had become their own mirror image, it would be weird but not problematic.
TRUE / FALSE

TEN THINGS A GENIUS KNOWS

① How to make a Möbius strip
The simplest non-orientable surface is called a Möbius strip, named after the German mathematician August Ferdinand Möbius. A Möbius strip is made by taking a long rectangle of paper (or any other flexible material) and joining the two short ends together with a half-twist. This half-twist creates the wondrous properties of the Möbius strip, because it glues the 'back' of the strip to the 'front', creating a surface that has only one side. The Möbius strip also has only has a single edge, in contrast to a cylinder that has two circular edges. The properties of the Möbius strip that are interesting to us are topological, in that they do not depend on the size of the paper but on the intrinsic structure of the object.

② What makes a Möbius strip non-orientable
A non-orientable shape is one in which notions of left and right (or clockwise and anticlockwise) are not consistent as one travels around the shape. The twist in the Möbius strip means that, if a creature lived in the surface (not *on* the surface), then as it travelled around the strip it would return home as the mirror image of itself. The creature itself would not think that it had changed, but rather that everyone else had become mirrored. In some sense, the Möbius strip is the source of all non-orientability, because any other non-orientable surface will have a Möbius strip within it.

③ The weird things a Möbius strip can do
Cutting up a Möbius strip in various ways can lead to surprising and magical results. A line drawn along the centre of the strip will seem to go around the strip twice before getting back to the beginning, and cutting along this line will give – not two smaller Möbius strips as one might expect – but a single larger loop with four half-twists in it. (This is because cutting in half creates a shape that is the same as the boundary, or edge, of the original object, and for a Möbius strip this is a single loop.) But cutting this new, four-twisted strip in half again does not give yet another larger loop with more twists, but two loops of interlinked paper. Another interesting cut of the original Möbius strip is a cut one-third of the way in from the edge: this again creates two interlinked loops, one of which is a Möbius strip and the other of which has four half-twists.

④ Variations on the Möbius strip
There are actually many ways of gluing together two ends of a rectangle to create a one-sided surface. First of all, there is a choice in the direction of the twist: it may be clockwise or anticlockwise, and these create two different Möbius strips (although they are homeomorphic). Another choice is in the number of half-twists that are made before the ends of the rectangle are joined. Any odd number of half-twists will give a surface that has only one side and that is intrinsically the same as a regular Möbius strip – a creature living in such a surface could not tell how many twists had been put in it. But if a Möbius strip with three half-twists is cut in half along the centre, the result will be a loop that is magically tied into an overhand knot.

⑤ Where Möbius strips are found
Möbius strips seem like an abstract mathematical curiosity, but they have found uses in the real world, too. Conveyer belts that introduce a twist wear evenly on both sides and so need to be replaced less often. Cowls are often knitted as Möbius strips so that they lie flat when they are worn. In high-frequency electrical circuits, a Möbius-shaped resistor can resist the flow of electricity while cancelling its own inductance, meaning it does not cause any magnetic interference. The double-helix twisting of DNA creates a multiply-twisted (orientable) Möbius strip when DNA is in a circular form, requiring special enzymes to deal with topological problems in its replication.

⑥ What a projective plane is
We can use Möbius strips to build more complicated non-orientable surfaces. Two Möbius strips can be glued together along their edges to create a new shape called a Klein bottle (below), named after the German mathematician Felix Klein. It is impossible to do the gluing without the resulting

shape intersecting itself – at least, in three dimensions. The Klein bottle is a shape whose natural home is in 4-D space, and it is a closed shape that has no inside or outside, in the same way that the Möbius strip has no front or back. A more obscure non-orientable surface than a Klein bottle is the real projective plane. It can be made by gluing a disc to the circular boundary of a Möbius strip, or by gluing the edge of a Möbius strip to itself with a twist. Yet another way of creating a projective plane is to take a hemisphere and glue together every pair of opposite points around the rim. Like the Klein bottle, the resulting surface will intersect itself in three dimensions and will have no inside or outside. It relates to projective geometry, where the basic objects are straight lines through the origin: points on opposite sides of a sphere lie on the same straight line; hence, why they are 'glued' together in the projective plane. (See page 73.)

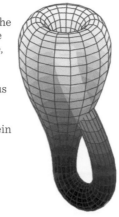

7 How to make a non-orientable shape in three dimensions
A Möbius strip may seem 3-D, but it is intrinsically 2-D. A creature living in the strip can only move forwards and backwards and from side to side, but it cannot move up and down, so it has only two degrees of freedom. Is it possible to make a 3-D object that is non-orientable? In the same way that a Möbius strip is made by joining two edges of a rectangle with a twist, we can glue opposite faces of a cube together with a twist. That is, glue the front face of a cube to the back face, but join the left edge to the right and the right edge to the left. If a person were then to walk through the front of the cube they would reappear from the back but mirrored. Other pairs of faces could also be joined to build up a complicated collection of non-orientable 3-D shapes!

8 How to define orientation in higher dimensions
For a manifold of any dimension, there is always a way of defining the orientation at a point. If the manifold is 1-D (a curve), then at each point we can define left and right. If the manifold is 2-D (a surface), then at each point we can define clockwise or anticlockwise. If the manifold is 3-D (a shape with volume), then we can choose either a right-handed or left-handed coordinate system. This pattern continues to higher dimensions, always allowing us a choice of two ways to orient space at a point. A manifold, like the Earth, can be represented by a collection of maps together with information on the overlaps between pairs of maps. To say that a manifold is orientable means that a choice of orientation has been made for each map, and that whenever a point is on two different maps at the same time, then the same choice of orientation is made on each one.

9 Whether our universe is orientable
Is it possible that our own universe could be non-orientable? The short answer is yes: while scientists believe that it is unlikely, nobody has definitively ruled out such a possibility, and many papers have been written exploring the consequences of a non-orientable universe. There are different kinds of non-orientability: space-non-orientable, time-non-orientable and space-time-non-orientable. One way the non-orientability could be introduced is through a wormhole that joins together two regions of space with a mirror-inducing twist. Particles travelling through the wormhole could emerge as their antiparticles (which have an interpretation as regular particles travelling backwards in time), or as their mirror images. Mirror particles are even potential candidates for dark matter.

10 Whether humans could survive in a non-orientable world
If the universe were space-non-orientable then a space traveller could go on a long journey and return to find that everybody and everything were mirrored. Initially, this might seem not to be a big problem, but many essential molecules, such as vitamins, sugars, hormones and amino acids, have a 'chirality' that means they would not interact with the body in the same way if they were mirrored. Most DNA also has the same handedness (a right-handed helix), so a mirrored person would not be able to reproduce with a non-mirrored person.

TALK LIKE A GENIUS

❡ Möbius was not the first person to write about the Möbius strip; this honour went to the German mathematician Johann Benedict Listing, the same man who coined the word 'topology'. ❡

❡ DNA does not naturally form non-orientable Möbius strips, but in 2010 scientists found a way to create a true DNA Möbius strip using a technique called DNA origami. They were also able to perform operations such as cutting it in half to produce a four-twisted strip. These techniques are a new tool in molecular engineering and could be used to make even more complex nanostructures in the future. ❡

❡ The symbol for recycling is usually represented as a three-twisted Möbius strip, as is the symbol for Google Drive. ❡

❡ "Situs inversus" is a medical condition in which a person's major organs are found on the mirrored side of their body; for example, the heart being on the right-hand side. Such a person has not travelled around a non-orientable universe because their basic molecules have not been mirrored, and so they are able to live normally with their strange condition usually undiagnosed. ❡

1 TRUE – It is made from a strip of paper by gluing two ends together with a twist, joining the 'front' to the 'back'.

2 TRUE – Cutting a Möbius strip in half gives one double-length strip with extra twists in it.

3 TRUE – A full twist is the same as two half-twists, and adding these to a strip of paper gives an orientable shape that becomes two interlocked loops when cut in half.

4 FALSE – The Klein bottle is an example of a closed surface where inside and outside are the same.

5 FALSE – We rely on chiral molecules to survive, so a person who became mirrored would not live for very long.

THE BLUFFER'S SUMMARY

In a non-orientable universe, a traveller could go off in one direction and return home to find that everything had become its mirror image.

Euler's formula and the shape of surfaces

'It is clear that there is no classification of the universe that is not arbitrary and full of conjectures.'

JORGE LUIS BORGES

The great Swiss mathematician Leonhard Euler noticed a curious pattern among the five Platonic solids, relating together how many corners, edges and faces were in a shape. The same formula worked for some other shapes, but not all, leading mathematicians to question what it was that Euler's formula was really measuring. The answer led to deep work in topology and a complete understanding of the theory of 2-D shapes, but left open a tantalizing question about shapes in higher dimensions.

From knowing only how many corners, edges and faces a shape has, could you draw it? With a few caveats, mathematicians have worked out that the answer to this question is 'yes'.

ARE YOU A GENIUS

1 A square-based pyramid has five vertices, eight edges and four faces.

TRUE / FALSE

2 For any shape, adding the number of vertices and faces, and subtracting the number of edges, always gives the same answer.

TRUE / FALSE

3 A line stretching to infinity in one direction is mathematically the same as a line stretching to infinity in both directions.

TRUE / FALSE

4 The count of vertices, edges and faces in an (orientable closed) surface can tell us how many holes the surface has.

TRUE / FALSE

5 The number of mathematically different surfaces is so vast that it is impossible to enumerate or make sense of them.

TRUE / FALSE

TEN THINGS A GENIUS KNOWS

① What Euler's polyhedron formula says

A polyhedron is any 3-D shape made up of flat faces, straight edges and sharp corners (called vertices). For example, a cube has eight vertices, twelve edges and six faces. A square-based pyramid has five vertices, eight edges and five faces. The dodecahedron, one of the five Platonic solids, has 12 pentagonal faces, 30 edges and 20 vertices. Euler noticed a pattern with these numbers: if we add up the number of vertices and faces of a polyhedron, and subtract the number of edges, then the answer always seems to be 2. The cube gives $8 - 12 + 6 = 2$; the pyramid has $5 - 8 + 5 = 2$; the dodecahedron has $20 - 30 + 12 = 2$. Euler's formula is written as $V - E + F = 2$.

② When Euler's formula is valid

Does Euler's formula work for *all* polyhedra? It does for all the Platonic solids, but it is easy to think of some examples where it fails. If we take a cube but remove one of the faces, for example, then we have $V - E + F = 1$. A humble square, with four vertices, four edges and one face also has $V - E + F = 1$. A more interesting example of a failure is the shape pictured below. It has 24 vertices, 48 edges and 24 faces, making $V - E + F = 24 - 48 + 24 = 0$. Why do some shapes fail the formula while others work? The answer relates to the topology of the shape.

③ That topologically equivalent shapes have the same Euler formula

In Euler's formula, we are simply counting things. If one of the faces of a shape bulged out a bit, or if an edge were curved, it would not change the count of how many there were. So faces and edges of a shape can be smoothly deformed without changing the shape's Euler formula, so long as different faces or edges do not intersect during the deformation. Dividing up the faces, edges and vertices of an existing shape also does not change the formula: splitting an edge into two smaller edges also creates a new vertex, and splitting a face

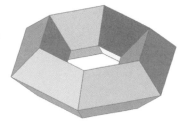

into two smaller faces creates a new edge, so overall $V - E + F$ remains the same. This type of argument can be made rigorous to show that topologically equivalent shapes always have the same Euler formula. The cube, pyramid and dodecahedron all have an Euler formula of 2 because they are all equivalent to spheres.

④ What it means to classify objects

Mathematicians love to classify things; that is, to put things into categories so that all examples of an object fall into exactly one category. They will then look for ways to decide which category an object is in. In topology, the categories are usually 'shapes that are topologically equivalent'. A connected (that is, all one piece) 1-D manifold is equivalent to either: an infinite line stretching in two directions, an infinite line stretching in one direction, a finite line, or a circle. Which of the four categories an object falls into is determined by asking two questions: is the object finite in length (if yes, it is a finite line or a circle; if no, it is a half-infinite or fully infinite line), and does the object have end-points (if yes, it is a half-infinite or finite line; if no, it is a fully infinite line or a circle). We say that these two questions 'solve the classification problem' for 1-D connected manifolds.

⑤ The classification problem for surfaces

In the same way that two questions completely determine the shape of a 1-D connected manifold, we can ask how many questions determine the shape of a 2-D connected manifold. Some examples of 2-D manifolds are: a sphere, a hemisphere, a torus (doughnut), a Möbius strip, a Klein bottle and a projective plane. Are these all topologically different, or are any of them equivalent? If we are presented with a complicated 2-D shape (such as the surface of the human body), how can we tell which category of manifold it falls into? This is called the classification problem for surfaces.

⑥ That there are only two families of surfaces

When classifying surfaces, mathematicians restrict themselves to those that are finite in size, connected and closed – the last feature meaning that the shape has no boundary. So a hemisphere is not a closed surface because it has a circular boundary, but a sphere is closed. A Möbius strip is not closed, but Klein bottles and projective planes are. Given this restriction, it turns out that surfaces fall into two families: those that are orientable and those that are not. Within each family, surfaces can be told apart by their Euler formula, and this provides a complete classification of surfaces.

7 **How the Euler formula classifies surfaces**
The family of orientable surfaces is determined by the number of holes in the shape, called the genus. A shape with g holes has an Euler formula $2 - 2g$. So a sphere (no holes) has Euler formula 2, while the torus or doughnut (one hole) has Euler formula 0. The shape shown below, therefore, has a Euler formula of -2, while if somebody told us a mystery orientable shape with an Euler formula of -6, we would know it was a surface with four holes in it. The non-orientable family is characterized by how many Möbius strips are glued into each shape. When there are k Möbius strips, the shape will have Euler formula $2 - k$. The projective plane (Euler formula 1) is created from a single Möbius strip, while the Klein bottle (Euler formula 0) is made by gluing together two Möbius strips.

8 **What the Euler formula is in higher dimensions**
As the Euler formula is so useful in helping us to distinguish different surfaces, perhaps it can be made to work for higher dimensions too, and in particular for 3-D shapes. The Euler formula is keeping count of the different ingredients that make up a shape: its 0-dimensional (0-D) components (vertices), 1-D components (edges) and 2-D components (faces). In three dimensions, we also need to count how many 3-D components it has. If this last count is given the letter M, the Euler formula in three dimensions would be $V - E + F - M$. Continuing the pattern, we can add on how many 4-D components are in a shape, subtract the number of 5-D components and so on. This more general formula for a shape is called the Euler characteristic.

9 **Whether the Euler formula can classify 3-D manifolds**
We have seen that in one dimension there is only one finite connected closed manifold: the circle. In two dimensions, there are two families of such manifolds: orientable and non-orientable, with the Euler formula being able to tell these apart. Is it possible to use the more general Euler characteristic to classify 3-D manifolds? Sadly not. It turns out that any closed 3-D manifold has an Euler characteristic of 0, so this calculation gives us no information at all (although it provides an interesting constraint on what 3-D manifolds can look like). This is true in any odd dimension, so the Euler characteristic is useless at telling apart manifolds in dimensions 5, 7, 9 . . . as well.

10 **Whether manifolds can be classified at all in high dimensions**
The problem of whether or not 3-D manifolds could be neatly classified in the same way as 2-D manifolds turned out to be one of mathematics' trickiest questions, taking mathematicians most of the 20th century to resolve. Its resolution involved proving the Poincaré conjecture (see page 104), one of the Millennium Problems worth one million dollars. Beyond that, the situation is hopeless. It has been shown that a classification of 4-D manifolds (and above) is impossible, because of an undecidable problem in group theory (see page 116) called the 'word problem'.

TALK LIKE A GENIUS

❛ A football is made up of 12 pentagons and 20 hexagons. Using Euler's formula it turns out that any football made of pentagons and hexagons will always have exactly 12 pentagons in it, though there is no limit to how many hexagons can be in the ball. Carbon molecules called fullerenes have this structure, and some have been found with more than 200 hexagons. ❜

❛ A heated debate among topologists is what kind of surface the human body is. This comes down to asking how many 'holes' are in a body. The digestive tract clearly creates one hole, which would give the answer that the human body is a torus. Another hole connects the nose (with two nostrils) and mouth, bringing the genus up to three. Of course, you can increase your genus to whatever number you like by getting piercings. ❜

❛ The Euler formula being non-zero for a sphere has a consequence called the Hairy Ball Theorem. This says that a hairy sphere can never be combed completely flat – there will always be a cowlick or tuft of hair that refuses to lie flat. The torus, though, does have Euler formula of 0, so a hairy doughnut can be combed flat. ❜

WERE YOU A GENIUS?

❙ FALSE – It has five vertices, eight edges and five faces.

❷ FALSE – Shapes that are topologically different will give different values for this answer.

❸ FALSE – These are different types of 1-D manifolds.

❹ TRUE – For a closed orientable surface, there is an exact relationship between the number of holes and the Euler formula.

❺ FALSE – The classification of surfaces shows that closed surfaces fall into just two categories, with shapes in each category enumerated by the Euler formula.

THE BLUFFER'S SUMMARY

Euler's formula relates the count of vertices, edges and faces in a shape to the number of holes in the shape, giving mathematicians a complete understanding of 2-D surfaces.

Identifying shapes

'The spheres of truth are less transparent than those of illusion.'

L.E.J. BROUWER

In the quest to find invariants that could distinguish between different topological objects, mathematicians invented the concepts of homotopy and homology. One is easy to define but hard to calculate, while the other is hard to define but easier to calculate. Both are methods to identify the different 'holes' in an object. Homotopy and homology became the foundation of algebraic topology in the 20th century and are becoming an important tool for data analysis in the 21st century.

Topological holes get far more complicated than the holes in your socks. To understand them, we need to bring in some heavy ideas from algebra and geometry.

1 The letter X can be smoothly deformed down to a single point.

TRUE / FALSE

2 Topologically, there is only one way to walk around the block and return home.

TRUE / FALSE

3 A torus, or doughnut, has two different kinds of hole in it.

TRUE / FALSE

4 It is possible to draw a knot on the surface of a doughnut, but not on the surface of a sphere.

TRUE / FALSE

5 In some strange universes it is possible to walk once around a circle and not be back where you started.

TRUE / FALSE

TEN THINGS A GENIUS KNOWS

1 What homotopy equivalence is

Two topological spaces are considered equivalent, or homeomorphic, if they can smoothly and reversibly be deformed into one another. The key here is the reversibility. We can define a weaker notion of topological equivalence, called homotopy equivalence, which does not worry about this aspect of things. For example, the letters A and P can be smoothly deformed into the letter O, but the letter O cannot be smoothly deformed back into either A or P. We say that A, P and O are homotopic but not homeomorphic. Another example of spaces that are homotopic, but not homeomorphic, are the Möbius strip and the circle: the width of the strip may be decreased to zero to make it into a circle.

2 The importance of homotopy equivalence of loops

Tie a piece of string to your front door. Walk down the road to a café, letting the string out after you. Buy a coffee, then walk back to your house. If, when you get home, you can pull the string back without it catching on anything, then the path you took was homotopically trivial (or contractible). That is, you can smoothly deform your path down to nothing. If, however, you walked around a tree or around the block on your way home, then you cannot pull the string back and you have walked a non-trivial path. The different types of loop we can walk tells us about the topology of our space.

3 What the fundamental group of a space is

The presence of a single non-trivial loop tells us that there is a hole in a space. Taking this further, we can try to draw all possible loops in a space, starting and ending at the same point, to see how many different holes there might be and how they relate to each other. We only care about paths that are homotopically different, so if one path can be smoothly deformed into another, they are considered to be the same. The collection of all such loops forms a mathematical group (see page 117) called the fundamental group. Hence, the rich structure of group theory can be used to answer questions about topology.

4 The fundamental group of the circle

If you walk around the block to get your coffee, your piece of string will trace out a non-trivial loop. But you could also walk around the block twice before coming home, and this traces out a different loop. Then again, if you walk around the block twice clockwise and once anticlockwise, two of the loops will cancel out and your total trip would be equivalent to a single clockwise loop. Each loop you make is measured by adding 1 for each loop clockwise and subtracting 1 for each loop anticlockwise. The collection of all possible loops in this space (which is a square, and so topologically a circle) is then equivalent to the set of integers. This is the fundamental group of a circle.

5 More interesting fundamental groups

In a torus, or doughnut, loops can be made to do more interesting things than in a circle. In the picture below, the loop travels around the torus many times before going back to the beginning, in a way that cannot be captured by a single number. There are actually two holes in a torus: one going around the large hole in the centre, and one going around the body of the torus. The pictured loop can be thought of as a loop that goes around the 'large' hole three times and the 'body' hole five times. In fact, any loop drawn on a torus can be decomposed in this way so that every loop is associated with a pair of integers. The fundamental group of the torus is then two copies of the integers. Not all fundamental groups are as simple as collections of numbers though. Removing a knot from space leaves a shape where the loops are highly intricate, giving a fundamental group that can only be described by complex algebra.

6 What the higher homotopy groups are

Any loop that is drawn on the surface of a sphere can be pulled tight to a point, which means that all loops on a sphere are trivial. A sphere is not a trivial topological object, but the fundamental group cannot detect this. The fundamental group is looking for all the 1-D holes in a shape; to detect higher-dimensional holes, we need to use higher-dimensional loops, by which we mean higher dimensional spheres. (A circle can be thought of as a 1-D sphere.) The nth homotopy group of a shape captures all ways of mapping an n-dimensional sphere into the shape, starting at the same point.

7 Why homotopy groups are so hard to calculate

Homotopy groups are easy for mathematicians to define, but notoriously difficult to calculate. One of the problems is that it is difficult to chop up a space into smaller pieces and relate the (higher) homotopy groups of the whole object to the groups of the small pieces. For example, a sphere is made by gluing together two hemispheres, but we cannot calculate the homotopy groups of a sphere by knowing the groups of the two hemispheres. In fact, although spheres are some of the simplest topological objects, there is no formula that tells us what the homotopy groups will be, not even in two dimensions.

8 The idea behind homology groups

Homology is a different approach to finding the various 'holes' in a space, and it does build up its knowledge of a manifold by looking in turn at the vertices, edges, faces, volumes and so on. A 1-D hole is a loop – that is, a cycle of edges that begins and ends at the same vertex. In the diagram, d is a loop, as is $a - b$ (edges are given a direction, so $-b$ means go backwards along the edge b). The path $b - c$ would also be a loop, but the presence of the face F means that the edges b and c are homotopy equivalent and can be deformed into each other. This means that when we look for holes, we ignore loops that are the boundaries of higher-dimensional faces.

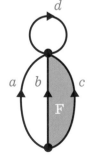

If two faces instead of one were attached between b and c, then these would create a 2-D hole, which would in turn be ignored if the space in between were filled in by a 3-D volume.

9 Examples of homology calculations

As an example of how homology works, and how strange it can be, consider the Klein bottle (see page 94). This shape can be represented by a square in which one pair of opposite sides (labelled a) are joined together with a twist, and the other pair (labelled b) are joined together without a twist. The four corners of the square all end up being joined to a single point, so they are all given the same label, V. The inside of the square is the single face of the Klein bottle,

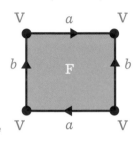

called F. We see that the edges a and b are both loops, since they join the vertex V to itself. We also know that the loop around the boundary of F should be considered as the zero loop. This loop is $a - b + a + b = 2a$. So we have found a loop a; if we go around it twice it is the same as doing nothing! This odd fact is used in physics to model fermions (such as electrons) that can have half-integer spin.

10 Some modern applications of homology

The way that homology is calculated from vertices, edges and faces makes it an excellent way to analyse information that is built up of data points. New techniques of topological data analysis look at finding the shape of data, and are particularly good for complex high-dimensional data that is difficult to analyse by other means. Edges are drawn between data points if they are close together (and the experimenter can decide what 'close together' means). One thing homology can easily do is tell when the data points are forming clusters – this was used in 2011 to detect a new form of breast cancer. Homology can also detect holes, such as gaps in the network coverage of dynamically changing Wi-Fi hotspots.

TALK LIKE A GENIUS

❝ Take a piece of paper, crumple it up, then put it back where it was on the table. Brouwer's fixed point theorem says that there will always be a molecule in the paper that ends up sitting directly above where it started. This surprising fact was proved using homology theory. ❞

❝ One consequence of homotopy theory is that, on the surface of the Earth, there are always a pair of antipodal points having the same temperature and pressure. This is a result of the Borsuk-Ulam theorem, another consequence of which is that we can put two pieces of bread and a piece of ham anywhere in the universe, and bisect them all with a single cut. ❞

❝ The plate trick demonstrates the idea that some loops are only zero if you go around them twice instead of once. Balance a plate on the palm of your hand, facing up. Now rotate your wrist clockwise, keeping the plate facing up. A rotation through 360° returns the plate to its initial position, but your arm will be twisted. Continue turning your wrist through another full rotation, and now both arm and plate are back to how they started. ❞

THE BLUFFER'S SUMMARY

Homotopy and homology apply the powerful techniques of group theory to the study of topological shapes, revealing the structure of different types of holes and providing new ways to analyse complex data.

The Poincaré conjecture

'Emptiness is everywhere and it can be calculated, which gives us a great opportunity. I know how to control the universe.'

GRIGORI PERELMAN

If a shape looks like a sphere, acts like a sphere and moves like a sphere, is it really a sphere? The Poincaré conjecture addresses this question in three dimensions, which surprisingly turns out to be the most difficult dimension of them all to understand. The problem was so difficult that it was listed as one of the seven Millennium Prize Problems by the Clay Institute, each worth one million dollars. In 2003, it became the first (and only) Millennium Problem to be solved, with the Russian mathematician Grigori Perelman providing the proof and solving a much bigger problem about the geometry of 3-D spaces.

Homotopy theory is one of the most powerful tools we have for understanding the structure of shapes, but is it even good enough to identify the humble sphere?

1 A beach ball is an example of a 3-D sphere.

TRUE / FALSE

2 There is only one way (topologically) to draw a circle on a beach ball, assuming the circle cannot intersect itself.

TRUE / FALSE

3 Spheres of any dimension contain a 1-D hole (that is, a hole that can be detected by drawing a circle around it).

TRUE / FALSE

4 The Poincaré conjecture was solved by finding a way to smooth out shapes so that spheres could be identified.

TRUE / FALSE

5 A 3-D manifold can be broken into pieces that each have one of eight different kinds of geometry.

TRUE / FALSE

TEN THINGS A GENIUS KNOWS

1 What a sphere is

There are many different ways to think about what a sphere is, and what it looks like in different dimensions. The most intuitive way to think of a sphere is as the collection of points that are all a fixed distance away from the origin. If our space is 1-D (picture the number line), then a sphere consists of two points: one each side of the origin. In a 2-D plane, the set of points a fixed distance from the origin is a circle. In three dimensions we get the surface of a ball, and so on. The intrinsic dimension of the sphere is one dimension lower than the space it lives in: two points is 0-D, a circle is 1-D and the surface of a ball is 2-D. When mathematicians talk about the dimension of a sphere, they always mean the intrinsic dimension, so a 2-D sphere is the surface of a ball. This means that when we talk about the 3-D sphere, we actually mean a sphere that is living in 4-D space.

2 How homology and homotopy analyse a sphere

If an object looks like a sphere, is it really a sphere? The answer will depend on the lens through which we are looking at the object. To a topologist, the lenses are usually homology and homotopy: two powerful tools for deciding when objects are topologically the same, each analysing the different types of 'hole' in a shape. Homology looks for voids in each dimension, while homotopy looks at the ways that spheres of different dimensions can be drawn in the object, which is a more subtle way of looking for holes. When the object is a sphere of dimension n, both methods come to the same conclusion: there is a single n-dimensional hole and no smaller holes. But is the converse true? That is, if a shape has no holes except for a single one in dimension n, is it an n-dimensional sphere?

3 The problem of homology spheres

When the French mathematician Henri Poincaré first developed his ideas on homology and homotopy, he believed that homology was powerful enough to detect spheres: in particular, the 3-D sphere. He soon realized he was wrong and constructed an example to show it. This example is called the Poincaré homology sphere, and is also known as Poincaré dodecahedral space, because it is constructed from gluing together sides of a regular dodecahedron, a shape with 12 pentagonal faces. Opposite faces are glued together with a slight twist (to get them to line up), and this creates a closed shape with exactly the same homology as a sphere. That is, it has no 1-D or 2-D holes, but it has a 3-D void. To detect that it was not a sphere, Poincaré realized that he needed to use homotopy.

4 What the Poincaré conjecture says

Homology is not good enough to detect spheres, but what about homotopy? In the case of the Poincaré homology sphere, its fundamental group (which is the first homotopy group, measuring the ways that a circle can be drawn on an object) can already detect that it is not a true sphere. This motivated Poincaré to pose a new question: will the fundamental group always detect when a (closed, finite) 3-D manifold is not a sphere? He conjectured that the answer was 'yes', and this became known as the Poincaré conjecture. So if a dodgy character tries to sell you a fake sphere, you can always detect the fakeness by finding a circle on the sphere that is caught around a 'hole' and cannot be shrunk to a point. If no such circle exists, the object really is a sphere. In a way, the result is surprising because it conjectures that a 3-D sphere can be detected using an intrinsically 1-D tool.

5 How the Poincaré conjecture works in higher dimensions

In higher dimensions it is not true that a shape with trivial fundamental group is necessarily a sphere, so a stronger condition needs to be met. The generalized Poincaré conjecture says that if the homotopy groups (in all dimensions) of a closed manifold are the same as a sphere, then the shape is topologically equivalent (homeomorphic) to the sphere. This sounds like a strong restriction for a space to have, but the conjecture is still surprising because, in general, homotopy is a much weaker kind of equivalence than homeomorphism (see page 101). Furthermore, since the conjecture was so difficult to prove in dimension three, it was thought to be completely intractable in higher dimensions. But, by 1962, Stephen Smale had a proof that the conjecture was true for all dimensions greater than, or equal to, five. In 1982, Michael Freedman solved the case for dimension four. (Both Smale and Freedman won Fields Medals for their work.)

6 **How Ricci flow changes a manifold**
In 1982, Richard Hamilton designed a new technique that he hoped would be able to prove the Poincaré conjecture. The technique was called Ricci flow, and involved putting a Riemannian metric on a manifold (see page 78) and then smoothing it out to see if the manifold would become a sphere. Ricci flow will contract areas with positive curvature and expand areas with negative curvature. Since a sphere has overall positive curvature, this means Ricci flow should end up shrinking spheres down to points in a finite amount of time. If Hamilton could show that a shape with trivial fundamental group could always be shrunk down to a point in this way, this would solve the Poincaré conjecture because, in the moment before its collapse, the shape would be a true sphere.

7 **The difficulties with Ricci flow, and Perelman's solution**
The reason Hamilton was not able to prove the Poincaré conjecture using Ricci flow was that, sometimes, the flow would get stuck. That is, it would hit what is called a singularity. Perelman showed that these singularities all had a simple form, which meant that he could cut the manifold along the singularities and continue the Ricci flow along the resulting pieces. Equally importantly, he showed that only a finite number of cuts were necessary to get down to a collection of pieces that were all spherical. From this point, he could glue the pieces together with cylinders, reconstructing the original manifold and showing that it was homeomorphic to a sphere.

8 **What the geometrization theorem says**
Perelman actually proved something stronger than the Poincaré conjecture, and this is a deep and far-reaching result called Thurston's geometrization theorem. Any 3-D manifold can be broken down in a standard way to a collection of smaller manifolds, and each of these manifolds has exactly one type of geometry. The result is analogous to the 2-D case, where surfaces have either Euclidean, spherical or hyperbolic geometry, except that in the 3-D case there are eight different possibilities for what the geometry can be.

9 **What it means for manifolds to be diffeomorphic**
By changing what we mean by 'looks like a sphere', variations on the Poincaré conjecture can lead to more interesting mathematics. The standard conjecture examines conditions that make a shape homeomorphic to a sphere, which is the usual notion of topological equivalence. Some manifolds come with a differentiable structure, where calculus can be done in a smooth way across the entire shape. In that case, two manifolds are diffeomorphic if they are homeomorphic *and* have the same differentiable structure.

10 **What the smooth Poincaré conjecture is**
Are there manifolds that are homeomorphic to a sphere, but not diffeomorphic to a sphere? In dimensions two, three, five and six, the answer is 'no', but in 1956 John Milnor discovered 'exotic spheres' in seven dimensions where the answer was 'yes'. These exotic spheres have the same intrinsic shape as normal spheres, but when it comes to doing calculus on them, there are additional strange and non-standard ways of doing so. It is still an open question whether exotic spheres exist in four dimensions. The conjecture that they do not is called the smooth Poincaré conjecture, though a number of mathematicians believe it will turn out to be false. It is possible that exotic spheres will turn out to provide a new way of looking at questions in particle physics and cosmology.

TALK LIKE A GENIUS

❦ Perelman was offered the Fields Medal in 2006 for his work, but turned it down stating, "I'm not interested in money or fame; I don't want to be on display like an animal in a zoo." In doing so, he became the first person to decline a Fields Medal. He also refused the one-million-dollar Millennium Prize by the Clay Mathematics Institute on the grounds that Hamilton had contributed at least as much as he had. ❧

❦ The work on the Poincaré conjecture shows how 3-D and 4-D space is inherently different to that of any other dimension. The techniques needed to prove the conjecture in these dimensions is completely different to those required for higher dimensions. In some sense, four dimensions is a kind of goldilocks dimension: not so big that anything can happen, not so small that nothing can happen, but just right for incredible amounts of complexity. Is it any wonder that we live in 4-D spacetime? ❧

❦ In 2003, a group of French and American scientists conjectured that the universe could be shaped like Poincaré's dodecahedral space. They looked at data from the cosmic microwave background radiation and found that it disagreed with predictions from the Standard Model at wide angles, bigger than 60°, but fitted with predictions of dodecahedral space. The theory does not have wide support among physicists. ❧

WERE YOU A GENIUS?

1 FALSE – The surface of a beach ball is a 2-D sphere; a 3-D sphere naturally lives in 4-D space.

2 TRUE – All circles can be deformed into one another on a 2-D sphere.

3 FALSE – Spheres of two dimensions and above have no 1-D holes: all circles drawn on spheres will deform down to a single point.

4 TRUE – Perelman solved the conjecture by using a method called Ricci flow to smooth out manifolds.

5 TRUE – This is Thurston's geometrization conjecture, proven true by Perelman's result.

THE BLUFFER'S SUMMARY

The Poincaré conjecture claims that if circles can be drawn on a finite 3-D shape without ever catching on any holes, then the shape is topologically a sphere. As of 2017 it is the only Millennium Prize Problem to have been solved.

The Hodge conjecture

'Algebra is nothing more than geometry, in words; geometry is nothing more than algebra, in pictures.'

SOPHIE GERMAIN (ATTRIB.)

The Hodge conjecture is one of the seven Millennium Prize Problems worth one million dollars, and is arguably the most difficult to understand. It analyses the structure of shapes described by simple equations – polynomials – but that live in spaces governed by complex numbers. If true, it would reveal a deep connection between algebra, geometry and calculus, with applications to number theory and fundamental physics.

While 20th-century mathematics has been successful in uniting algebra and geometry, it will need a 21st-century mathematician to get to the heart of the connection.

ARE YOU A GENIUS

1 Straight lines and circles can be described as the solutions of polynomial equations.

TRUE / FALSE

2 Working with equations using complex-numbered coordinates instead of real numbers doubles the dimension of the resulting shape.

TRUE / FALSE

3 There are six ways of writing the number 5 as the sum of two (non-negative) whole numbers.

TRUE / FALSE

4 String theory predicts that we live in a universe with a large number of dimensions, with most of them curled up so small that we can't see them.

TRUE / FALSE

5 Soap bubbles always take the shape that maximizes the surface area they take up

TRUE / FALSE

TEN THINGS A GENIUS KNOWS

❶ What an algebraic variety is
Many of the shapes familiar to us can be
described as the solutions of a polynomial equation.
For example, a circle of radius 1 is the set of all
coordinates (x,y) that solve the equation $x^2 + y^2 = 1$. We
are familiar with a straight line being described as $y = mx + c$. A sphere has equation $x^2 + y^2 + z^2 = 1$. In general,
a shape that can be described as the set of solutions
to a polynomial equation (or equations) is called an
algebraic variety.

❷ The difficulty of finding subvarieties
Hodge theory is concerned with algebraic
varieties that are also manifolds – that is, shapes
that look locally like our standard Euclidean space.
Within a manifold, we can find smaller manifolds
called submanifolds. For example, a sphere is a
manifold, and the equator of a sphere – a circle – is
also a manifold. In the same way, we can ask whether
a shape within a variety is itself a variety. This turns
out to be an extremely subtle and difficult question
despite its apparent simplicity, and is the question
that the Hodge conjecture tries to answer.

**❸ How complex numbers make
varieties complicated**
A polynomial equation like $x^2 + y^2 = 1$ can be defined
over the real numbers or it can be defined over the
complex numbers. The latter would mean that x
and y would each have the form $a + ib$, where i is
the square root of -1 and a and b are real numbers.
Mathematicians like to
work over the complex
numbers because the
fundamental theorem
of algebra certifies
that every polynomial
equation will have a
solution. But working
over the complex
numbers doubles
the dimension because each coordinate itself has
two dimensions. Elliptic curves, which are very
important in cryptography, are algebraic varieties
that have the shape of a torus when defined over the
complex numbers. They have submanifolds that are
circles, but these cannot be subvarieties, because

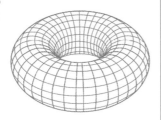

circles are intrinsically 1-D while complex varieties
must have an even-numbered dimension.

**❹ Why mathematicians use algebra to
understand geometry**
Geometry is, in general, considered to be difficult,
while algebra is, in general, straightforward.
Mathematicians therefore try to translate
geometrical problems into algebraic problems so as
to make them easier to solve. An analogy is a doctor
using a machine like an MRI scanner to see inside
the body of a patient. Algebraic topology uses the
machines of homology and homotopy to see inside
manifolds and identify submanifolds. Algebraic
geometers want a machine to see inside varieties,
and one that is powerful enough to identify the
subvarieties. The machine used by Hodge theory is
called cohomology.

❺ What cohomology is
Cohomology studies functions on
submanifolds, concentrating on certain special
functions that give the same output when applied
to topologically equivalent submanifolds. Different
types of function produce different types of
cohomology. Singular cohomology looks at linear
maps that input submanifolds and output rational
numbers (integer fractions). De Rham cohomology
uses the techniques of calculus and looks at
functions involving k-forms. These are infinitesimal
units of length, area, volume and so on. A k-form
integrated over a k-dimensional submanifold will
output a number that is the total length/area/
volume/and so on, of that submanifold. If the
manifold in question is defined over the complex
numbers, the outputs of these integrals will be
complex numbers.

**❻ How Hodge theory refines a
cohomology group**
Manifolds that are complex algebraic varieties
have a lot of structure. Hodge theory exploits this
structure to add more information into the de
Rham cohomology groups. It shows how to split up
k-forms into different categories based on
how the calculus interacts with the complex
structure. A 1-form splits up into (1,0) and (0,1)
forms, while a 2-form splits up into (2,0), (1,1) and
(0,2) forms. There are as many splittings as there

are ways of writing k as the sum of two non-negative integers (so 2 can be written as $2 + 0$, $1 + 1$ or $0 + 2$). The cohomology group corresponding to k-forms splits up along the same lines, and this refinement of the group provides a way of recognizing subvarieties. A subvariety whose dimension is p less than the dimension of the whole variety will correspond to an element within the (p,p) de Rham cohomology class.

7 **What the Hodge conjecture says**
In the MRI-scanning analogy, the doctor knows how to start with a body and produce a scanned image. But if they start with a printed image, can they decide whether it is the scan of a real body? Similarly, mathematicians know how to start with a subvariety and find its image within de Rham cohomology – it will be in the (p,p) class. But they want to be able to start with an element in the (p,p) class of cohomology and decide whether it comes from a subvariety. Doing this would give mathematicians a machine for determining the general structure of varieties. The Hodge conjecture posits an answer to this question by combining the complex-valued de Rham cohomology with the rational-valued singular cohomology. It theorizes that, whenever an element is both in the (p,p) de Rham cohomology group and also in the $2p$-dimensional singular cohomology group, then it should come from a linear combination of subvarieties.

8 **What evidence there is for the Hodge conjecture**
In 1924, over ten years before Hodge announced his conjecture, Solomon Lefschetz proved a result called the theorem on (1,1)-classes. It says that the Hodge conjecture is true for $p = 1$, meaning it is true when the subvarieties have a dimension that is one less than the dimension of the whole variety. Allied with another result, called the hard Lefschetz theorem, we know that the Hodge conjecture is true whenever we start with a variety of at most three (complex) dimensions. In particular, it is true for an especially beautiful class of manifolds called K3 surfaces.

9 **What a K3 surface is**
K3 surfaces are a very special class of complex 2-D varieties (meaning they are 4-D in regular coordinates). They are like 4-D soap bubbles, in that they are shapes that minimize their surface area. Simple enough to study, yet complicated enough to be interesting, they are ubiquitous in string theory as they are examples of Calabi-Yau manifolds. Traditional models of string theory predict that the universe has ten dimensions, with six of them curled up so small that we cannot detect them. But only certain kinds of shapes have the right kind of geometry to curl up so small and to keep the symmetries we see around us. Calabi-Yau manifolds have exactly the right kind of geometry for this to happen, and Hodge theory is essential for understanding them.

10 **Why we care about the Hodge conjecture**
The Hodge conjecture seems incredibly obscure, impossible to be stated in any kind of simple terms. Few results beyond Lefschetz's theorems give evidence that it is true, apart from a small number of special cases. For a random algebraic variety, there are no subvarieties to care about. But for special classes of varieties, the Hodge conjecture makes deep predictions about the possible shapes that the subvarieties can have. Such interesting cases arise often in number theory, and in particular Diophantine equations, where Hodge theory has helped to prove some of the most profound theorems we know. Hodge theory unites many areas of mathematics in unexpected ways – topology, geometry, Galois theory, complex and rational numbers, calculus and group theory. Its proof will not only be worth one million dollars but will create incredible new insights into mathematics.

TALK LIKE A GENIUS

❝ Solomon Leftschetz was born to a Russian Jewish family in 1884 and originally trained to be an engineer, moving to the United States to work for the Westinghouse Electric Company in Pittsburgh. However, tragedy struck in 1907 when he lost both hands in an accident. After recovering, he took up mathematics and became a giant in the field of topology. Each morning, a student had to slide a piece of chalk into his artificial hand, and each evening they had to remove it. It was said of Lefschetz that he never wrote a correct proof or stated an incorrect theorem. ❞

❝ K3 surfaces were named by André Weil after three algebraic geometers and a mountain: Kummer, Kähler and Kodaira, along with the mountain K2 in Kashmir. Weil is unlikely to have climbed K2, but spent two years in India early in his career and took a vacation to Kashmir. The trip left him with "a prodigious treasure trove of impressions" that stayed with him for the rest of his life. ❞

❝ Mathematics, like any science, is often a case of trial and error. Hodge's initial conjecture was disproved by Atiyah and Hirzebruch, leading to a new conjecture, which is the one worth one million dollars today. Hodge then made a more general conjecture, which was disproved by Alexander Grothendieck in a paper humiliatingly titled: "Hodge's general conjecture is false for trivial reasons". ❞

WERE YOU A GENIUS?

1 TRUE – They are examples of algebraic varieties.

2 TRUE – Each complex coordinate itself has two dimensions, which doubles the overall dimension of the shape.

3 TRUE – We can write 5 as 0 + 5, 1 + 4, 2 + 3, 3 + 2, 4 + 1 and 5 + 0.

4 TRUE – Traditional models of string theory predict that our universe has ten dimensions, with six of them curled up very small. Hodge theory helps us understand the types of shapes that have the right geometry to do this.

5 FALSE – Soap bubbles always have minimal surface area. Hodge theory can help us to understand 4-D soap bubbles called K3 surfaces.

THE BLUFFER'S SUMMARY

The Hodge conjecture is one of the most complicated pieces of mathematics there is, but it is basically a prediction that will help us understand the structure of complex shapes.

Knot theory

'Knots are more numerous than the stars; and equally mysterious and beautiful . . . '

JOHN TURNER

Knots have existed in human culture for thousands of years. They are indispensable to sailors, climbers and dressmakers, and they are seen decorating everything from religious texts to jewellery and monuments. Since the 19th century, they have also found their way into mathematics, where they form important examples of topological shapes. Simple questions, such as whether the left-handed and right-handed overhand knots are different, turned out to be surprisingly difficult to answer and required new genius ideas to solve them. After a century of studying these beautiful shapes in an abstract way, mathematicians are now working with scientists to put them back into practice.

Simple enough to draw and describe, knots also have a potentially infinite complexity, making them fascinating objects of study for mathematicians, scientists and artists alike.

1 A circle is not a mathematical knot because it isn't tangled.

TRUE / FALSE

2 The reef knot (right-over-left, then left-over-right) and granny knot (right-over-left, then right-over-left) are the same knot.

TRUE / FALSE

3 It is possible to untangle a knot by tying a second knot in the same piece of string.

TRUE / FALSE

4 If a length of string (like your headphones) is agitated in a bag, the most likely knot to form is the trefoil (overhand) knot.

TRUE / FALSE

5 DNA can become knotted in your cells.

TRUE / FALSE

TEN THINGS A GENIUS KNOWS

1 **What a mathematical knot is**

A topologist can always untangle a piece of string, no matter how complicated its tangle might be, because the free ends can be manipulated bit by bit to get rid of the knot. So, to be able to study knots mathematically, the two ends of the string must be fused together to trap the knot in the string. Once the ends are fused, the knot cannot be untangled without cutting the string. the string may be stretched and moved about, but the structure of the knot within it will not change. Intrinsically, the string has the shape of a simple circle (which is 1-D), but the way that this circle is placed in 3-D space provides infinitely many possibilities for different tangles. The intrinsic simplicity compared with the extrinsic complications make knots a very interesting area of study.

2 **How mathematicians draw knots**

Knots are represented using 2-D drawings, which means that care must be taken to accurately capture the 3-D information of how the knot twists around itself. The four pictures below are examples of knot diagrams, in which strands of the knot are only allowed to meet at crossings where one strand goes completely over or under another. More than two strands cannot meet at once. There are many diagrams for the same knot, depending on the angle from which the knot is viewed, or how the tangles are arranged.

3 **The basic question in knot theory**

Given two circular pieces of string, the basic question in knot theory is: are they the same knot? That is, can one piece of string be moved and manipulated (without cutting or gluing) to look like the other one? Even more fundamentally, how can we tell whether a tangled piece of string is knotted at all? In the four pictures shown above, the first and third knots look different but are actually the same: they are right-handed trefoil (overhand) knots. The fourth picture shows a string that can be unravelled

to form a circle that is not tangled at all, and this is called the unknot. The second picture looks like the first, except that it has been mirrored. This is called the left-handed trefoil, and it was an open question for many years whether it was truly different from the right-handed trefoil.

4 **What a knot invariant is**

The main tool that knot theorists use to tell knots apart is the idea of a 'knot invariant'. This is some property of a knot that does not change when the knot is bent, stretched or twisted. In particular, if an invariant is computed from a knot diagram, then it should give the same answer if it is computed using two different diagrams of the same knot. Counting the number of times the knot crosses over or under itself is not an invariant, because the diagram of the unknot above would give an answer of three, but the unknot can also be drawn with no crossings at all. However, the *smallest* number of crossings a knot can be drawn with *is* an invariant, called the crossing number. The unknot has crossing number 0 while the trefoil has crossing number 3, proving that they are different.

5 **What the unknotting number is**

When two knots give different values for an invariant, this confirms that they are distinct knots. If the invariant gives the same value then this does not show that they are the *same*, however. Like a suspect for a crime, if they do not match the witness's description then they cannot be the perpetrator; if they do match it does not mean they are guilty. For example, the two knots here both have a crossing number of 5, but they are topologically different. They can be distinguished using a different invariant called the unknotting number, which is the number of times the string must pass through itself before

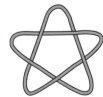

it becomes unknotted. The knot in the top picture on the previous page can be unknotted in a single move but the knot in the bottom picture requires at least two moves.

6 **More sophisticated knot invariants**
The unknotting number is an invariant that is easy to state, but notoriously hard to calculate, because every possible sequence of unknotting steps must be tested for every possible diagram of a knot. It cannot be done by computer because there are infinitely many diagrams to check. There are other invariants that sound more complicated, but are easier to calculate. Knot polynomials are some of the most common knot invariants. These are algebraic expressions, such as $t - 1 - t^{-1}$, that can be algorithmically computed from a knot diagram and are generally very sensitive at telling apart knots. The Alexander polynomial was the first to be discovered in 1928, but the most famous is probably the Jones polynomial, the invention of which won Vaughan Jones the Fields Medal in 1990 and which provided the first simple proof that the left- and right-handed trefoils were distinct.

7 **The difference between prime and composite knots**
Just as numbers can be either prime or composite, so, too, can knots. A prime knot is one that cannot be created by tying two smaller knots in a piece of string. For example, the trefoil and the two 5-crossing knots above are prime knots, but the reef knot is not prime because it is made by tying two overhand knots in the same piece of string. (The granny knot is also made by tying two overhand knots in a piece of string, but it uses two of the same trefoil while the reef knot uses one left-handed trefoil and one right-handed trefoil.)

8 **How many knots there are**
Since Victorian times, mathematicians have been making tables of prime knots, organized by crossing number. The number of prime knots of each crossing number increases exponentially, so that there are only two knots of crossing number 5, 165 knots of crossing number 10, and nearly 10,000 knots of crossing number 13. Tables of knots are currently only complete up to crossing number 16, and are extremely difficult to make, since we know of no general algorithm to tell knots apart. There are infinitely many knots, so the tables will never be complete, but the more knots we know of, the more examples there are to test new theories and invariants.

9 **Some open questions in knot theory**
The main open question in knot theory is finding an invariant, or collection of invariants, that can recognize when a piece of string is unknotted. The Jones polynomial is a potential candidate, though mathematicians mostly think this is unlikely. Some methods do exist for recognizing the unknot, but they are too complex to be of practical use. (A fast algorithm for unknot recognition would also solve the P vs NP problem – see page 176). Another problem is being able to tell when a knot and its mirror image are distinct, as in the case of the trefoil. It is still unknown whether, if a knot is made up of tying two smaller knots in the same piece of string, the crossing number of the new knot is the sum of the crossing numbers of the two smaller knots.

10 **How knot theory is used to analyse DNA**
DNA is like a two-metre-long piece of string found inside each of the cells in our bodies. Since a cell can be up to ten times smaller than the width of a human hair, the DNA must go through a process called supercoiling to be able to fit into the cell. This can be problematic when your body is trying to access the information in the DNA – for example, if it is trying to grow new cells. As well as the problems caused by supercoiling, the double-helix shape of the DNA results in it becoming knotted around itself when it replicates. Special enzymes called topoisomerases have the ability to fix these problems by performing unknotting operations, and knot theorists are working with biologists to try to understand how they work. Such research has become unexpectedly important in developing cancer drugs because inhibiting topoisomerase can stop tumour cells from growing.

TALK LIKE A GENIUS

❝ The serious study of knot theory as a mathematical discipline began with a completely wrong theory of science. In 1867, Lord Kelvin had a theory that atoms were knotted vortices of aether, and that different knots would account for the different elements in the periodic table. Though the theory was nonsense, his friend Peter Guthrie Tait dedicated his life to tabulating knots to explore the theory and he created knot invariants that are still being used today. ❞

❝ The Jones polynomial has a connection with Chern-Simons theory, which, in turn, can be used to model quantum phenomena and other theories of physics, such as string theory and supergravity. In particular, the Jones polynomial may help scientists to develop a quantum computer based on topology. ❞

❝ Scientists now have the ability to create knotted molecules, which is exciting because every different kind of knot could yield a new and interesting material. They might make new bullet-proof armour, non-stick frying pans or smart nanomaterials that could mend themselves. ❞

1 FALSE – A mathematical knot is any circle living in 3-D space, be it tangled or untangled.

2 FALSE – These are different because right-over-left and left-over-right are mathematically different knots. Reef knots are much stronger in your shoelaces than granny knots!

3 FALSE – It is not possible to tie two knots in the same piece of string that 'cancel' each other out.

4 TRUE – This result won an Ig Nobel prize in 2008, and showed that the knots that formed from the string depended on the length and flexibility of the string, with the lowest crossing number knots being the most likely to form.

5 TRUE – Special enzymes are needed to unknot DNA in order to keep you alive.

THE BLUFFER'S SUMMARY

Telling mathematical knots apart is a notoriously difficult problem, with solutions now important for analysing the DNA in our cells.

The mathematics of symmetry

'The universe is built on a plan the profound symmetry of which is somehow present in the inner structure of our intellect.'

PAUL VALERY

The world is full of symmetry, from the shape of the human body to the petals in a flower or the patterns on your wallpaper. There are symmetries in time, such as your daily routine or the chiming of a clock. And there is symmetry in the numbers on the number line and in pebbles laid out on a beach. In the 19th and 20th centuries, mathematicians distilled the idea of symmetry down to its bare basics and discovered amazing structures hidden within the everyday symmetries we take for granted. The new subject was called group theory and it has now become one of the foundations of mathematics and physics.

In what ways can we change an object and yet leave it looking the same? Simple ideas about symmetry have evolved into the intricate and important field of group theory.

1 A symmetry of an object means a reflection or a rotation.

TRUE / FALSE

2 Rotating a square and then reflecting it (leaving it looking the same) always has the same effect as doing a single reflection.

TRUE / FALSE

3 If you have four different objects, there are 24 ways to lay them out in a row.

TRUE / FALSE

4 With three pebbles in a row, swapping the first and second and then the first and third would give the same ordering as swapping the first and third and then first and second.

TRUE / FALSE

5 The laws of physics change depending on where in the universe we are.

TRUE / FALSE

TEN THINGS A GENIUS KNOWS

1 What symmetry is to a mathematician
The everyday notion of symmetry, as in a face or a flower, suggests an object that looks the same in a mirror or when rotated. Mathematicians think of a symmetry as being an action: something one can do to an object to leave it looking the same as it did before. The two ideas are often related. For example, a square has eight different symmetries based on geometry: we can rotate it by 0°, 90°, 180°, and 270°, or we can reflect it in half horizontally, vertically or along the two diagonals.

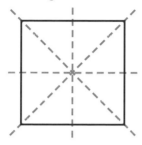

2 How a symmetry can be non-geometrical
Not every symmetry comes from rotations and reflections. A different example of a symmetry comes from the idea of permutations. Picture four identical pebbles arranged in a line. Any rearrangement of these pebbles is a symmetry, because the pebbles cannot be told apart. There are 24 possible rearrangements, so there are 24 possible symmetries of the pebbles. Permutation symmetries form one of the most important classes of symmetries, and are used to study roots of polynomials as well as puzzles like the Rubik's cube.

3 How symmetries behave
Two symmetries of an object can be combined one after the other to form a new symmetry. For example, rotating a square through 90° and then reflecting it in the horizontal axis has the same effect as a single reflection along the top-right to bottom-left diagonal. Another simple observation is that every object has at least one symmetry: the action that does nothing. Finally, it is always possible to undo a symmetry by applying another symmetry. A 90° rotation of a square can be undone by a 270° rotation, and a horizontal reflection can be undone by applying a second horizontal reflection.

4 The definition of a group
Group theory is the abstract study of symmetry. A group is a set of objects, S ("the symmetries"), that can be combined together in a way that resembles how symmetries behave. Firstly, combining any two elements of S must result in another element of S. Second, the set S must have an element called the identity, which does nothing when combined with any other element. Third, every element of S has an inverse, which undoes the action of the element to give the identity back again. There is a final technical fourth rule, called associativity, which says that $a*(b*c) = (a*b)*c$ for any a, b and c in S, where $*$ denotes the method of combining elements.

5 How numbers can become groups
The number line can be made into a group in different ways, highlighting the structure of different types of numbers. The integers have a shift-symmetry: moving the line a whole number left or right leaves the line looking the same. This is the same as saying that the integers are a group under the operation of addition. Adding two whole numbers gives another whole number. The identity is 0, and the inverse of any number is its negative. But the integers are not a group under multiplication. Multiplication stretches (or compresses) the number line, and stretching the integers changes how they look. We can multiply two integers to get another integer, but we cannot go back again: dividing one integer by another does not always give a whole number. However, the rational numbers and the real numbers (without 0) are both groups under multiplication, using 1 as the identity and division as inverses.

6 What an abelian group is
In some groups, the order of operations matters. Flipping a square horizontally and then rotating it by 90° gives a different answer from first rotating and then flipping. Swapping the first two pebbles in a line and then swapping the first and third gives a different answer from performing the swaps in the reverse order. But in other groups we can perform the operations in any order we like, such as when we are adding integers or multiplying fractions. If, in the group, we have $a*b =$

$b*a$ for any a and b in S, then the group is called abelian, named after the Norwegian mathematician Niels Abel.

7 How to spot groups inside groups

In the example of the symmetries of a square, the collection of rotations forms a group by itself. Combining any two rotations creates another rotation, and the identity element is a rotation by 0°. When a subset of a set S forms a group in its own right, under the same operation * as the original group, then this subset is said to create a subgroup. A further subgroup of the square is the one that consists of just two rotations: 0° and 180°. In the permutation group of pebbles, an example of a subgroup is the group consisting of those permutations that leave the fourth pebble fixed and just rearrange the first three.

8 What a cyclic group is

The rotations of a square are an example of a cyclic group. This means that all of the symmetries can be generated by just one symmetry, taking that symmetry and performing it over and over again. In this case, rotation by 90° is the generator. Performing it twice gets us the 180° rotation, three times gives us the 270° and four times gives us the 0° rotation. This is a cyclic group of order four, because the generator must be applied to itself four times to get all the elements of the group. The group consisting of the identity and the horizontal flip of the square is a cyclic group of order 2. But the entire group of symmetries of the square is not cyclic, because no single symmetry can generate all eight actions. The integers under addition is a cyclic group of infinite order, generated by the element 1 and its inverse -1.

9 How to tell when two groups are the same

Sometimes two different-looking groups turn out to have the same structure. The symmetries of an equilateral triangle have the same structure as the permutations of three pebbles. There are six symmetries of a triangle: three rotations (by 0°, 120° and 240°) and three reflections. There are also six ways of permuting three pebbles. To see that the two groups are 'the same', label the three corners of the triangle as A, B and C and consider these to be the three pebbles. Every permutation of the pebbles corresponds to a symmetry of the triangle: for example, swapping A and B corresponds to a

reflection in the line through C, while moving A to B, B to C and C to A corresponds to a rotation through 120°. When the structures of two groups are the same, they are called homomorphic.

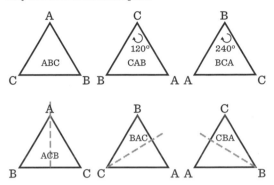

10 Why group theory is important in science

In chemistry, the study of crystals relies on group theory. Crystallography uses a particular group called the space group, which describes all the possible symmetries that a three-dimensional (3-D) lattice could have – rotations, reflections, translations, screw axes and glide planes (see page 81.) This group has 230 elements in it, each one describing a different type of crystal that can exist in real life. Group theory is also at the heart of modern physics. The mathematician Emmy Noether proved in 1915 that, whenever there is a symmetry of a system, there is a corresponding law of conservation. For example, our physical laws are true regardless of where in the universe we are, or at what point in time, and these symmetries give us conservation of momentum and mass, respectively. Symmetries are thus one of the keys to understanding the universe in which we live.

TALK LIKE A GENIUS

◖ The set of all possible moves that can be made on a Rubik's cube forms a group, where the group operation is to do the moves one after the other. This group has over 43 billion billion elements in it, and it is non-abelian, making it an immensely complicated structure. Yet it was proved in 2010 that it is possible to solve the cube from any configuration using only 20 moves. ◗

◖ You may have come across the 15 puzzle, which is a 4 × 4 grid of squares forming a picture, with one square removed in the bottom-right corner so that the pieces can move. The puzzle is to rearrange the squares and see if you can get them back into the right order. To play a practical joke on your friend, take the puzzle apart and put it back together with two adjacent squares swapped. Group theory can prove that the puzzle will now be impossible to solve! ◗

◖ Group theory made an appearance in the animated series *Futurama*, in the 10th episode called 'The Prisoner of Benda'. A machine allows the characters to switch minds with each other, but no pair of people can ever switch minds twice. Is it possible for everyone to get their own minds back in their own bodies? The creator of the show, Ken Keeler, actually proved a new result in group theory in order to resolve the plot line. ◗

1 FALSE – A symmetry is any action done to an object that leaves it looking the same.

2 TRUE – This can be verified by checking all possible combinations of reflections and rotations.

3 TRUE – There are four options for laying down the first object, then once this is chosen there are three options for the second, then two for the third and one for the fourth, giving a total of 4 × 3 × 2 × 1 = 24 choices altogether.

4 FALSE – If the three pebbles are labelled A, B, C, then the first/second followed by first/third gives C, A, B, while the other way gives B, C, A.

5 FALSE – The laws of physics are the same everywhere in the universe, and this symmetry gives us the law of conservation of momentum.

THE BLUFFER'S SUMMARY

Group theory is the abstract study of symmetry and is the key to understanding the structure of the number line, the laws of physics and the study of crystals.

The impossibility of solving equations

'I've said it before: equations are the devil's sentences. The worst one is that quadratic equation, an infernal salad of numbers, letters and symbols.'

STEPHEN COLBERT

The story of solving polynomial equations is full of tragedy and genius. It involves two young men who died before they were 30, whose deaths could have been prevented if only their mathematical discoveries had been appreciated while they yet lived. Niels Abel and Evariste Galois both worked on one of the hardest problems of their time: why nobody could find a formula for the quintic equation. Getting to the heart of the problem required the invention of a completely new area of mathematics that is still being investigated today.

The quadratic formula we all learn in school was the stepping stone to a question that required a genius (or two) to solve it.

ARE YOU A GENIUS

1 The roots of a polynomial are those numbers that you input to give an answer of zero.
TRUE / FALSE

2 Some polynomial equations do not have any solutions at all.
TRUE / FALSE

3 Nobody could solve quadratic equations until the 16th century.
TRUE / FALSE

4 To get from PEST to STEP requires swapping four pairs of letters.
TRUE / FALSE

5 There are 60 ways of rotating a dodecahedron (a shape with 12 pentagonal faces) to leave it looking the same.
TRUE / FALSE

TEN THINGS A GENIUS KNOWS

1 What a polynomial is
I am thinking of a number. If I square it, minus five lots of the number, and add 6, the answer is 0. What number am I thinking of? This puzzle can be written as a polynomial equation: if the unknown number is written as x then the riddle asks you to solve $x^2 - 5x + 6 = 0$. This is called a quadratic equation because the highest power of x is 2. If the highest power of x were 3 it would be called a cubic equation, for 4 it would be a quartic and for 5 it would be a quintic. The values of x that solve the equation are called the roots of the polynomial, and the numbers on the left-hand side of the equation (in this case, -5 and 6) are called the coefficients.

2 How to solve a quadratic equation
Most people can probably still recite the formula for solving a quadratic equation: 'minus bee plus or minus the square root of bee-squared minus four ay cee over 2 ay'. But what does this mean? It means that if we have a quadratic equation $ax^2 + bx + c = 0$ then we can find the roots by using the formula

$$x = \frac{-b \pm \sqrt{b^2 - 4ac}}{2a}$$

For the riddle above, this formula tells us that x must be either 2 or 3. This is a formula that was known to ancient Greek and Indian mathematicians, and possibly also the Babylonians and Egyptians. But it took until the mid-16th century for people to find a corresponding formula for solving cubic equations – a formula where you could input the coefficients and it would output all the roots. Discovery of a formula for the quartic came soon after, but nobody could then find a formula for quintic equations.

3 What the Abel-Ruffini theorem says
Nobody could find a formula for the solutions of a quintic equation because such a thing does not exist. In essence, this is what the Abel-Ruffini theorem says. More precisely, it states that the roots of a quintic polynomial cannot have a formula given in terms of radicals: that is, using only addition, subtraction, multiplication, division, roots and powers. The Italian mathematician Paolo Ruffini,

in 1799, gave the first attempt at a proof, but it was incomplete and was turned into a correct proof by the Norwegian mathematician Niels Abel in 1824. The proof does not say that quintics do not have roots (the fundamental theorem of algebra tells us that every quintic has five roots, although some of them may be complex numbers) but only that the roots cannot always be expressed in a certain algebraic way.

4 Who Evariste Galois was
A general formula for the roots of a quintic was impossible, but certain polynomials did have 'nice' solutions that could be written in terms of radicals. Was there a way of deciding which polynomials were 'nice' and which were not? This was the problem that Évariste Galois solved during his short but brilliant life. Like Abel, Galois's achievements were not recognized during his lifetime. Rejected by the mathematical establishment and by Stéphanie, the woman he loved, Galois died by a pistol shot to the stomach after a duel over Stéphanie. The night before the duel, in 1832, he wrote up his mathematical ideas, which were published and gave rise to a whole new area of mathematics called Galois theory.

5 How group theory is involved in polynomials
Galois and Abel invented group theory, the cornerstone of modern algebra, in order to solve the problem of the quintic. Their insight was to look at the symmetries of the roots of polynomials and to explain how the symmetries of the roots could lead to a formula for solving the polynomial. The key was looking at all the different integer equations that the roots satisfied, and seeing which permutations of the roots kept the equations being true. For example, in the quartic polynomial $x^4 - 14x^2 + 9$, the roots are A = $\sqrt{2} + \sqrt{5}$, B = $\sqrt{2} - \sqrt{5}$, C = $-\sqrt{2} + \sqrt{5}$ and D = $\sqrt{2} - \sqrt{5}$, and we have that $(A + B)^2 = 8$, $A + D = 0$ and $AC = 3$. We cannot switch A and B and keep all the equations true, but we can switch A and B at the same time as C and D. This collection of permutations is called the Galois group of the polynomial.

6 How the Galois group reveals a formula for solutions
In the example above, the Galois group of the polynomial consists of the identity polynomial (which keeps all the roots the same) and three pairs of swaps: swapping (A,B) and (C,D), or (A,C)

and (B,D), or (A,D) and (B,C). Each of these permutations has order 2, meaning that if we do them twice it is the same as doing nothing. This fact tells us that the formula for the roots of this polynomial will involve square roots; and indeed the formula here is $x = \pm\sqrt{7 \pm 2\sqrt{10}}$. In other quartic equations the Galois group might have permutations of order three, in which case there would be cube roots, or permutations of order four, which would give us fourth roots.

7 How to decide if a permutation is even or odd

A permutation can be described as either even or odd, depending on how many 'swaps' are needed to create it. For example, the permutation above, where we switched A and B at the same time as switching C and D, is an even permutation because there are two swaps. The permutation that rearranges (A,B,C) to (B,C,A) doesn't look like it contains any swaps at all, but this rearrangement can be accomplished by doing two swaps, one after the other: first switch A and B to get (B,A,C), then switch A and C to get (B,C,A). This permutation is also even. However, the permutation taking (A,B,C,D) to (B,C,D,A) is odd because it takes three swaps.

8 What the alternating group is

Every permutation can be described as being either even or odd. The collection of even permutations on n objects forms a group in its own right, called the alternating group, and is written A_n. The group A_5 can be visualized geometrically as the group of rotations of an icosahedron or a dodecahedron (such as the one at top right). What Galois and Abel realized was that it was the structure of A_5 that was preventing quintic polynomials from being solvable.

9 Why quintic polynomials cannot be solved

We have seen that, when permutations of, for example, order 3 appear in the Galois group, this indicates that a cube root will appear in the formula for the solution of the polynomial. The full theorem, making this precise, says that if the Galois group of the polynomial breaks down into a collection of cyclic groups, then the polynomial has a formula for its roots. The surprising fact about the group of permutations on five elements is that it does *not* break down this

way: the subgroup A_5 of even permutations is a simple group and does not break down into cyclic groups. Furthermore, every finite group is the Galois group of some polynomial, so there will be a quintic polynomial with A_5 as its Galois group.

10 What the quintic polynomial led to

Abel, Ruffini and Galois had together solved the problem of why there was no formula for a general quintic, and moreover how to decide whether any particular polynomial had a formula for its roots. But their work was far more wide-reaching than this result. Mathematicians after Galois developed his work into an area called field theory, which looks at what happens when new numbers are added into old number systems. For example, complex numbers are created when the new number $\sqrt{-1}$ is added into the real numbers. The roots of the polynomial $x^4 - 14x^2 + 9$ can be analysed by adding the new numbers $\sqrt{2}$ and $\sqrt{5}$ to the rational numbers. This approach has yielded a rich and interesting theory of numbers that is used in cryptography and deep questions such as Fermat's Last Theorem.

TALK LIKE A GENIUS

❛ Abel's proof of the impossibility of solving the quintic went largely unappreciated during his lifetime, in part because he was so poor that he could only afford six pages of paper on which to write his proof, making it very difficult to read and understand. Had his work been acknowledged, he would have been famous, but unfortunate circumstances meant that he remained poor, contracting tuberculosis and dying at the age of 26, just two days before the offer came of a permanent academic position. ❜

❛ Galois always had a hot temper that got him into trouble. At the entrance examination for the École Polytechnique, the best university in France at the time, Galois got into an argument with his examiner and threw the blackboard eraser at him. (He did not get in to the university.) Later in his life he was arrested twice, the first time for threatening the king and the second time for leading an armed protest wearing a banned military uniform. His time in jail may have been productive, though, in giving him space to develop his mathematical ideas. ❜

1 TRUE – The roots of a polynomial are the numbers that give zero.

2 FALSE – Every polynomial of degree n has n roots, by the fundamental theorem of algebra (though sometimes these may be complex numbers).

3 FALSE – Quadratic equations could be solved by the ancient Egyptians, Babylonians and Greeks, as well as Indian and Persian mathematicians in the seventh and ninth centuries, respectively.

4 FALSE – It can be done in three swaps: PEST → SEPT → STPE → STEP.

5 TRUE   There are 24 symmetries through pairs of opposite faces, 15 symmetries through pairs of opposite edges, 20 symmetries through pairs of opposite vertices and one symmetry that does nothing.

THE BLUFFER'S SUMMARY

The mathematical study of symmetry was invented in order to prove that there could not be a general formula for solving polynomial equations.

The building blocks of symmetry

'All the moves we were making seemed to be forced. It was not perversity on our part, but the... nature of the problem that seemed to be controlling the directions of our efforts and shaping the techniques being developed.'

DANIEL GORENSTEIN

The classification of finite simple groups is one of the longest and most difficult results in mathematics, covering over 10,000 pages of journal articles, involving more than 100 different authors and taking over 100 years to complete. It is a testament to collaboration, perseverance, insight and intuition, and it provides the foundation for many other areas of mathematics.

Can you wrap your head around the longest proof in mathematics, including the Monster it contains?

1 Any collection of symmetries breaks down in a structured way into smaller symmetries; for example, symmetries of a square are reflections or rotations.

TRUE / FALSE

2 There is a shape that has more symmetries than any other shape (ignoring those that have infinitely many symmetries, such as the circle).

TRUE / FALSE

3 Any rotation symmetry of a regular hexagon can be obtained by rotations through 180° (order 2) and 120° (order 3).

TRUE / FALSE

4 It is possible for 15 schoolgirls to walk out in five groups of three, for seven days in a row, so that no pair of girls walks out in the same group more than once.

TRUE / FALSE

5 Mathematicians have a complete list of all the different types of finite symmetries that could possibly exist.

TRUE / FALSE

TEN THINGS A GENIUS KNOWS

① How to divide one group by another
Prime numbers form a 'classification' of numbers in the sense that every whole number can be written uniquely as a product of prime numbers. We can take any number and keep dividing it by primes until we get down to 1. When Evariste Galois developed the idea of groups in 1832, he tried to find a way to do a similar thing for groups, and find a way to 'divide' one group by another. His idea was that when we divide a group by a subgroup, the resulting object should form a group in its own right. It turns out that this is possible precisely when the subgroup has a property called 'normality'.

② What a normal subgroup is
If we take a group with eight elements and divide it by a group with four elements, our intuition dictates that the answer be a group with two elements. As an example, take the symmetries of a square, which is a group containing the identity, three rotations and four reflections. The three rotations plus the identity form a subgroup, and 'dividing' by this subgroup separates the symmetries into two: 'the rotations' and 'the reflections'. In this new two-element group the rotations are the identity, meaning that we can think of the rotations as 'doing nothing'. Reflecting, rotating and reflecting back is the same as a rotation. In general, this structure is what defines a 'normal subgroup'.

③ What a simple group is
A simple group is one that has no normal subgroups within it, except for the identity. If a group is not simple, it is possible to divide it into a collection of simple groups. First, find the biggest normal subgroup inside the group, then find the biggest normal subgroup inside that subgroup and so on, until no more subgroups can be found. This list of normal subgroups is called a composition series. At each stage in the process, dividing by the maximal normal subgroup gives a corresponding factor group, which is simple. In this way, the factor groups are like the prime numbers of a group: in some sense, they multiply together to give the group its structure.

④ Why the Jordan-Hölder theorem is important
There is only one way to write each composite number as a product of prime numbers, which is why prime numbers are considered the building blocks of the whole numbers. In order to think of simple groups as the building blocks of groups, we need a similar result that tells us that division can only be done in one way. But this isn't strictly true: in the example of the square there are two different normal subgroups, each with four elements (and these two subgroups have different group structures). Dividing each of these into smaller groups gives a total of seven ways of writing a composition series. However, the Jordan-Hölder theorem comes to the rescue, asserting that the collection of factor groups in each case is the same, up to permutation. So these factor groups, which are simple groups, can indeed be considered the building blocks of any group.

⑤ How it was possible for mathematicians to understand all simple groups
Just as there are infinitely many prime numbers, there are infinitely many finite simple groups. What was it that made mathematicians believe it was possible to understand them all? First, abelian groups (those in which the order of combining elements does not matter, so $a * b = b * a$) are easy to understand because every subgroup is a normal subgroup. Next, a big theorem called the Feit-Thompson theorem showed that any finite group with an odd number of elements could be broken down into cyclic groups. This meant that the classification problem was restricted to non-abelian groups of even order. But non-abelian even-order groups have involutions (elements of order two), and the Brauer-Fowler theorem had shown that there were only finitely many simple groups built using these involutions.

⑥ What the classification of finite simple groups says
From the proof of the Feit-Thompson theorem in 1963, mathematicians believed that a list of all finite simple groups would be theoretically achievable, although nobody believed it would be done in their lifetime. But, in 2004, the final piece of the puzzle was published and the list was complete. The result is called the 'classification of finite simple groups', and says that finite simple groups are either one of

a few different families of group, or they are one of 26 sporadic groups that do not seem to fit into any pattern or family. (The regular families consist of either prime-number-sized cyclic groups, certain permutation groups, or certain types of continuous symmetry groups called Lie groups.)

7. How to construct the Mathieu groups

The two smallest sporadic groups, called M_{11} and M_{12}, are named after Émile Mathieu, who discovered them in 1861, and contain 7,920 and 95,040 elements, respectively. The way they are constructed can be written as a puzzle: try to write down a collection of six-element sets – with the elements taken from $\{1, \ldots, 12\}$ – in such a way that no two sets have five elements in common, and so that every choice of five numbers from $\{1, \ldots, 12\}$ appears in at least one set. It turns out there is exactly one way to do this, and the symmetries of this collection of sets are M_{12}, with M_{11} the subgroup that fixes one point. Three more simple groups, M_{22}, M_{23} and M_{24} use numbers up to 24 to construct similar eight-element sets.

8. How a 24-dimensional lattice produces strange groups

In 24 dimensions, there is an unusually efficient way of packing spheres: this arrangement is called the Leech lattice. In such a lattice, each sphere can touch 196,560 neighbours. This configuration was so highly symmetric that John Leech wondered if any simple groups might be discovered within it, and John Conway agreed to look at the problem. Within a short time he found three sporadic simple groups within the symmetries of the Leech lattice, the largest of which had four million million million elements in it. These were called Co_1, Co_2 and Co_3. In addition, Conway rediscovered four other sporadic simple groups that other people had found and was able to use the Leech lattice to reconstruct the five Mathieu groups.

9. Just how big the Monster group is

The Monster group is the biggest of the sporadic groups and was predicted to exist in 1973 by Bernd Fischer and Robert Griess, although it took nearly ten years before Griess found a way to construct it. Fischer had created three new sporadic groups, Fi_{22}, Fi_{23} and Fi_{24} building from the Mathieu

groups. By generalizing his method further, he discovered a group called the Baby Monster with 4×10^{33} elements in. But this would turn out to be a subgroup of an even larger group called the Monster group, with 808,017,424,794,512,875,886,459,90 4,961, 710,757,005,754,368,000,000,000 different elements in it. Twenty of the 26 sporadic groups were contained inside the Monster; these were called the 'happy family', with the other six dubbed 'pariahs'.

10. The future of finite simple groups

Daniel Gorenstein, who initiated and oversaw the programme of classification, also started a programme of revisionism in which the immense first proof was simplified so that more people could understand it. The first attempt has seen the proof shrink from over 10,000 pages to about 5,000, but this is still far beyond any other proof seen in mathematics. As part of this work, mathematicians have tried to find an overarching theory that links all the finite simple groups together. One potential avenue concerns the monstrous moonshine theory, which tries to link the sporadic groups to modular functions through ideas in physics, including ideas in string theory and quantum gravity.

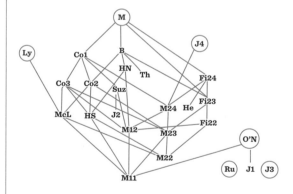

❦ The proof of the Feit-Thompson theorem was one of the longest that the mathematical community had ever seen, weighing in at 255 pages. But this feat would later be dwarfed by a proof by Aschenbacher that was 1221 pages long, which was the final puzzle piece in the classification of finite groups. Aschenbacher's proof is still one of the longest single papers ever written. ❧

❦ The order of the Monster group is divisible by the supersingular primes: 2, 3, 5, 7, 11, 13, 17, 19, 23, 29, 31, 41, 47, 59 and 71. These primes are related to supersingular elliptic curves, which are, in turn, important in cryptography and were mentioned by the character Charlie in the Season 2 episode 'In Plain Sight' of the television crime drama *NUMB3RS*.❧

❦ The *ATLAS of Finite Groups* is a book listing information on 93 of the finite simple groups, and was a project started by John Conway. Over time he acquired collaborators Curtis, Norton, Parker and Wilson, whose surnames all coincidentally have the same structure of consonants and vowels. This observation was noticed by the authors and is captured on the *ATLAS* itself with the names being written in an array that looks just like one of the character tables inside the book. ❧

1 FALSE – Some collections of symmetries are 'simple' and cannot be broken down into smaller symmetry types.

2 FALSE – We can find shapes with any finite number of symmetries (for example, rotation symmetries of a regular shape with n sides), so there is no biggest one.

3 TRUE – Rotation through $1/6$ of a circle, for example, can be obtained by two 120° rotations followed by a 180° rotation. The result is related to the fact that the prime factorization of 6 is 2×3.

4 TRUE – This puzzle was first posed by Thomas Kirkman in 1850 and has seven different solutions.

5 TRUE – The classification of finite simple groups provides a full list of finite simple groups, from which all other finite groups can be constructed.

THE BLUFFER'S SUMMARY

The classification of finite simple groups describes the structure of all finite symmetries and uncovers the mind-blowingly large and complex Monster group.

Continuous symmetry

'The most unexpected theories, from arithmetic to quantum physics, came to encircle this Lie field like a gigantic axis.'

JEAN DIEUDONNÉ

Groups were invented to help us understand the symmetries in the roots of an equation or the shape of a polygon. But arguably their most powerful influence has been in modern physics, where the notion of continuous symmetry is important – in the symmetries of time and space, for example. The objects built to study continuous symmetry are called Lie groups (pronounced 'Lee'), and though many families of Lie groups are well understood, there are a few rogue groups that are far more intricate and fascinating than anybody could have imagined.

When looking for symmetric shapes, forget squares, pentagons or hexagons. The infinite symmetries of the circle put every other shape to shame, and demand a new area of mathematics to deal with them.

1 There are shapes that have infinitely many symmetries.

TRUE / FALSE

2 A hairy ball always contains a cow-lick.

TRUE / FALSE

3 Electrons have a property called 'spin', which arises because electrons are modelled as balls spinning on an axis.

TRUE / FALSE

4 The Standard Model of physics, which models all elementary particles and forces, is explained using just three symmetry groups.

TRUE / FALSE

5 The symmetries of a certain 57-dimensional object account for the most complicated continuous symmetry group that we know of.

TRUE / FALSE

TEN THINGS A GENIUS KNOWS

1 What's special about a circle

The symmetries of a square, or of the whole numbers on the number line, are discrete. But the symmetries of a circle are continuous. Every single angle between 0 and 360° is a possible rotation symmetry and every line through the centre of the circle is a reflection symmetry. Picking a symmetry and changing it a tiny amount gives us a new symmetry. This is not true of the square: we may rotate by 90°, but we cannot rotate by 91° to leave the square looking the same. The circle also happens to be a manifold: a shape that looks like flat Euclidean space on small scales. Moreover, it is a differentiable manifold, which means that the manifold is created in such a way as to allow calculus to be done on it.

2 What a Lie group is

A Lie group is any object that is both a differentiable manifold and a group, with the requirement that the group operation must be compatible with the way that calculus is done on the manifold – its 'smooth structure'. The circle is one example of a Lie group. Another easy example is standard Euclidean space under the operation of addition, which corresponds to the symmetries of translation. The group consisting of all reflections and rotations of n-dimensional Euclidean space, leaving the origin fixed, is another Lie group, called the orthogonal group, O(n). The subgroup of O(n) consisting only of the rotations is also a Lie group, called the special orthogonal group SO(n).

3 Whether all spheres are Lie groups

The 1-D sphere – the circle – is a Lie group. Does this mean that all higher-dimensional spheres are also Lie groups? In two dimensions the sphere is indeed a differentiable manifold with a group structure, but the hairy ball theorem stops the group structure being compatible with the smooth structure. This theorem states that it is impossible to comb a hairy ball flat, and prevents 2-D spheres from being 'parallelizable' – a property that every Lie group has. An analogue of the hairy ball theorem applies to all higher dimensional spheres except for dimensions three and seven, and, of these two, only the 3-D sphere is a Lie group.

4 How Lie groups show up in physics

Lie groups are fundamental to our understanding of modern particle physics. This is due to a theorem by the German mathematician Emmy Noether, who showed that every continuous symmetry of a physical system has a corresponding conservation law. For example, the laws of physics are the same regardless of where in the universe we are, in which direction we are looking or what the time is. This gives us conservation of linear momentum, angular momentum and energy, respectively. Symmetries are important at the quantum level, too, controlling things such as the conservation of electric charge, the electroweak force and quark colours.

5 Which Lie groups are the most important

A particular collection of Lie groups, called the classical groups, are considered the deepest and most important part of the theory of continuous symmetries. The classical groups contain matrices (arrays of vectors) where the entries are either real numbers, complex numbers or quaternions. Complex numbers form a 2-D space: they have a real coordinate and an imaginary coordinate. Quaternions are 4-D: they have one real coordinate and three imaginary coordinates. The group SO(n) contains all rotations of real-valued vectors and corresponds to symmetries of spacetime. The group SU(n) is the subgroup of SO($2n$) that contains rotations of complex-valued vectors and is important for quantum phase space symmetries. Finally, Sp(n) is the subgroup of SU($2n$) dealing with rotations of quaternionic-valued vectors, which is important in Hamiltonian mechanics.

6 Which Lie groups are in the Standard Model of physics

The Standard Model of particle physics is based on a combination of three Lie groups: SU(3) × SU(2) × U(1). These groups have dimensions eight, three and one respectively, giving rise to the $8 + 3 + 1 = 12$ different elementary particles called fermions (quarks, neutrinos, electrons, muons and taus) and the 12 forces called bosons. The Lie group SU(3) describes the symmetries of the eight gluons and their interaction with quarks. The group SU(2), equivalent to a 3-D sphere, describes the symmetries of the three W and Z bosons that control the weak interaction responsible for radioactive decay. And U(1), a Lie group with the same structure as a circle, is responsible for the behaviour of photons, which carry the electromagnetic force.

7 How a Lie algebra relates to a Lie group
Every Lie group is associated with an object called its Lie algebra, which captures the idea of 'local' symmetries near the identity element. Formally, the Lie algebra is the space of vectors that are tangent to the Lie group at the identity. For example, at a point on a circle, the tangent is the straight line that just touches the circle, and so the Lie algebra of the circle corresponds to the real number line. It can happen that two different Lie groups have the same Lie algebra, so they have the same structure locally, but not globally. An important example is electron spin, which is modelled by the Lie group SU(2). This has the same Lie algebra as SO(3), the rotations in 3-D space, which is why it makes sense to think of an electron as a spinning ball.

8 How to classify Lie groups
Simple Lie groups are a special class of Lie groups that form the building blocks of other Lie groups, in analogy with the simple groups of general group theory. A Lie group is called simple if it is connected (a shape in one piece), non-abelian (so ab is not necessarily the same as ba) and contains no smaller connected subgroups. Simple Lie groups were classified by Wilhelm Killing in 1890. They turn out to be the four families of complex classical groups SU($n+1$), SO($2n+1$), Sp($2n$) and SO($2n$) (labelled A_n, B_n, C_n, and D_n, respectively) plus five exceptional groups E_6, E_7, E_8, F_4, and G_2.

9 Why E_8 is considered so special
The Lie group E_8 is the largest and most complicated of the exceptional Lie groups, consisting of the symmetries of a 57-dimensional shape. As a group, it has dimension 248. Inside any simple Lie algebra is a special collection of vectors called a root system, and the E_8 root system consists of 240 eight-dimensional vectors. The relationship between these vectors is encapsulated in the picture below. A particular set of eight of these vectors forms a lattice in 8-D space and this happens to correspond with the best possible way of packing spheres in eight dimensions. For a similar reason this lattice

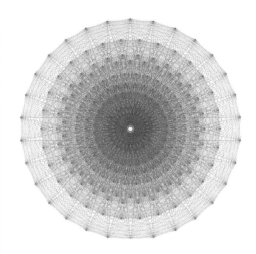

is an exceptionally efficient way of encrypting information. In 2007, a group of mathematicians finished an immense calculation of the structure of E_8, which contained more data than the Human Genome project.

10 Whether E_8 will provide a Theory of Everything
The classical Lie groups have formed the backbone of theoretical and particle physics for many years, leading some mathematicians to wonder whether the exceptional Lie groups might also have a role to play in explaining how the universe works. In 2007, the physicist Anthony Garrett Lisi proposed a unified theory of particle physics and gravity that tried to describe all particle interactions in terms of E_8 symmetries. His work has not been widely accepted by scientists, but E_8 does appear elsewhere in theories of everything. String theory makes particular use of the E_8 lattice in order to curl up 16 of the 26 dimensions in its model. And, in 2010, an experiment on electron spins in a crystal appeared to exhibit some of the E_8 symmetries.

TALK LIKE A GENIUS

❧ Noether was one of the leading mathematicians of her time, but as a woman her path was a difficult one. At the University of Erlangen she could sit in on lectures but not take the final exam. Even her attendance required the permission of the professor. In 1905, she was invited to apply for a position at the University of Göttingen by David Hilbert and Felix Klein, but there were protests by the faculty who did not want a woman there. Hilbert responded by saying 'I do not see that the sex of the candidate is an argument against her admission. After all, we are a university, not a bath house.❜

❧ Wilhelm Killing was a modest man who took holy orders so that he could get a teaching position in the town of Braunsberg. He invented many areas of mathematics independently of other people without getting the credit: he invented Lie algebras independently of Sophus Lie, described the Weyl group when Weyl was only three years old, and described the Coxeter transformation 19 years before Coxeter.❜

❧ The exceptional Lie group E_8 appears to have a link with the Monster group; an intriguing connection that is still being investigated.❜

THE BLUFFER'S SUMMARY

Lie groups capture the idea of continuous symmetry, which is what forms the foundation of our Standard Model of physics, explaining all the different particles and forces we see.

The Yang-Mills and mass gap problem

'If you are a researcher, you are trying to figure out what the question is as well as what the answer is.'

EDWARD WITTEN

The Yang-Mills and mass gap problem seeks a comprehensive mathematical foundation for a theory of physics that is our best current understanding of how particles work. It is part of the search for a 'theory of everything', uniting the long-range theory of relativity with the short-range theory of quantum mechanics. Quantum field theory has already made exceptional progress in describing the electromagnetic and weak forces. The hope is that, with the solution of this problem, which is one of the seven Millennium Prize Problems worth one million dollars, we will understand the strong nuclear force as well – the very force that holds matter together.

Our understanding of the fundamental forces of nature is mathematically incomplete, likely requiring genius ideas from many different angles.

1 A magnetic field describes the strength and direction of magnetic forces acting at each point of space.

TRUE / FALSE

2 The type of equations describing magnetic and electric fields are very similar to those describing the strong and weak nuclear forces holding atoms together.

TRUE / FALSE

3 Light sometimes behaves like a wave, and sometimes behaves like a stream of particles.

TRUE / FALSE

4 The Higgs boson, discovered in 2013, explains why certain particles have mass.

TRUE / FALSE

5 The quantum field theory that describes the strong nuclear force is called quantum electrodynamics.

TRUE / FALSE

TEN THINGS A GENIUS KNOWS

1 What a field is in physics

We have all come across fields in the school laboratory, in scattering iron filings around a magnet. The patterns and lines that the iron filings make show the underlying magnetic field: lines of force that, at each point, have a strength and directionality. The gravitational field is another example of a field. At each point in the universe, matter will be pulled in a particular direction by all the gravitational forces acting there. This pull will change in direction and intensity depending on where in space and time we are. Physicists think of fields as physical quantities occupying all of space, and they use field equations to describe what different fields look like at each point of spacetime.

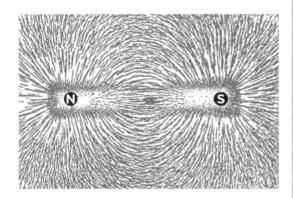

2 Whether there are fields for every force

Our current best understanding of physics says that there are four fundamental forces: electromagnetism, gravitation, the strong nuclear force (which holds together the nuclei of atoms) and the weak nuclear force (which is responsible for radioactive decay). One of the greatest triumphs of 19th-century physics was James Clerk Maxwell's unification of the electric and magnetic fields, and his discovery of the field equations for electromagnetism. We also have a good understanding of the gravitational field, thanks to Einstein's theory of relativity. There are classical fields for the strong and weak nuclear forces, but these do not accurately describe the physics we see. This is because these forces act over very small distances (as in an atomic nucleus) where strange quantum effects become important.

3 The counter-intuitive nature of quantum mechanics

Einstein won the Nobel Prize in 1921 not for his work on relativity, but for his discovery of a phenomenon called the photoelectric effect. Shining light onto different metals produced strange effects that indicated that light beams were composed of discrete packets of energy, which we now call photons. This was at odds with our understanding of light as a wave. It seemed that light acted as both a wave and a particle simultaneously, a property called wave-particle duality. Quantum mechanics proposed that the wavelike behaviour of light (and other fundamental particles like electrons) was due to the uncertainties inherent in matter: that position and velocity were not precise numbers, but were instead probabilities.

4 What quantum field theory is

The concept of wave-particle duality hinted at a link between field theory and quantum mechanics. Maxwell's theory of electromagnetism described light as a wave, while quantum mechanics described light in the language of discrete photons. It was Paul Dirac, in 1927, who found a way to quantize the electromagnetic field, creating a new subject that later became called quantum electrodynamics (QED). His idea was to picture a quantum field in a simplistic way as a network of tightly packed springs, each spring connected to its neighbours so that a disturbance of a particular spring would create a wave of oscillations moving through the network. Such a disturbance is called a photon. Since a minimum amount of energy is required to set a spring in motion, this would account for the quantum units of energy that we see. QED became the first example of a quantum field theory, or QFT.

5 The success and difficulties of QED

Dirac's creation of quantum electrodynamics was not the end of the story. Although the theory was put forward in the 1920s, it took until the 1950s before the mathematical foundations of the theory were put into place. Julian Schwinger, Richard Feynman and Shin'ichirō Tomonaga all won a Nobel

Prize for their work, the most important step of which was called renormalization, which removed troublesome infinities from the calculations. QED has gone on to be the most successful theory of physics ever created, making more accurate predictions than any other theory we have. It gave hope that the other three fundamental forces could also be understood by the mathematical ideas of quantum field theory.

6 How symmetry groups come into field theory

An important part of any field theory is the idea of a gauge symmetry. This is any transformation of the field that results in the same observed quantities. Einstein's field equations for relativity give the same answer after any change of coordinates (hence, the idea of everything being 'relative'), and the equations of QED give the same answer after any quantum change of phase. The symmetry group corresponding to this change of phase is the Lie group U(1) (see page 129). In order to make quantum field theory work for the other fundamental forces, physicists needed to identify the relevant symmetry groups (gauge groups) for each one.

7 What the Yang-Mills equations are

In 1954, the physicists Chen-Ning Yang and Robert Mills replaced the group U(1) of electromagnetism with a more general group. This new group was a 'compact simple Lie group', and was significantly harder to work with because it could be non-abelian. That is, combining symmetries in different orders might give different answers. The resulting Yang-Mills equations were so complicated that, even today, few people have found any exact solutions to them. But they provided a framework of how quantum field theories could be created for the other fundamental forces, once physicists had found the right gauge group for each one.

8 How Higgs saved the day

Yang-Mills theory was problematic because it predicted that force-carrying particles would have no mass. This was fine for photons, but false for the particles carrying the other forces. In 1967, physicists Sheldon Glashow, Abdus Salam and Steven Weinberg successfully created a QFT combining the weak nuclear force with electromagnetism by using the gauge group SU(2) × U(1) and introducing a new field to give force-carrying particles non-zero masses. Their new theory was called the electroweak theory and the new field was called the Higgs field. In 2013, the Higgs boson was discovered at CERN, leaving the success of the theory in little doubt.

9 What the QFT is for the strong force

Yang-Mills theory had another success when it came to explaining the strong nuclear force. Quantum chromodynamics (QCD) is a QFT built on the gauge group SU(3), describing the interactions of eight different force-carrying particles called gluons. These hold together quarks, which are the building blocks of protons and neutrons. QCD does not introduce a new field like the Higgs to get around the idea of massless particles, but instead uses a property of Yang-Mills theory itself, called 'asymptotic freedom', to explain why the strong force only acts at short ranges. All predictions made by QCD have turned out to be true, and yet the mathematical foundations of the theory are frustratingly shaky.

10 What the Yang-Mills and mass gap problem asks

There is currently a lot of experimental evidence that has not been predicted by Yang-Mills theory. It is not that Yang-Mills theory is wrong, but rather that it is incomplete. One of the seven Millennium Prize Problems asks mathematicians to construct a 4-D quantum Yang-Mills theory that explains why force-carrying particles have a non-zero mass. Gauge theory unites differential geometry with group theory and topology, and a solution to this problem will likely require a whole new area of mathematics that has not yet been developed.

❦ Edward Witten, so far the only physicist to have won a Fields Medal, is a pioneer of supersymmetric quantum field theory as well as string theory and quantum gravity. He is often named as the world's smartest living physicist and likely understands quantum field theory better than anyone else alive today. Yet even he is quoted as saying that QFT is "by far the most difficult theory in modern physics". ❧

❦ Nobody yet has any idea how to quantize gravity. Current field theories that try to combine the graviton (carrying the gravitational force) with the strong and electroweak interactions run into the same problems with infinities that plagued early attempts at quantizing electromagnetism. The ideas of string theory and loop quantum gravity are attempts to move away from field theories and find new explanations for how the universe might work. ❧

❦ Quantum field theory predicts that the vacuum of space is never truly empty. In the metaphor of the network of springs, each spring is continually in motion, and we see this in practice by the spontaneous creations of particle and anti-particle pairs. They are called "virtual" particles, but can affect the motion of real particles passing through. ❧

1 TRUE – The shape of the field can be made visible by scattering iron filings on a piece of paper sitting above a magnet.

2 FALSE – The strong and weak nuclear forces act on small scales, and so cannot be described by field theory without the introduction of quantum mechanics.

3 TRUE – QFT attempted to explain this wave-particle duality, treating photons (light particles) as waves travelling through a quantum field.

4 TRUE – Its discovery confirmed predictions of the electroweak theory.

5 FALSE – Quantum electrodynamics is the QFT for electromagnetism, while quantum chromodynamics is the QFT for the strong nuclear force.

THE BLUFFER'S SUMMARY

The Yang-Mills and mass gap problem asks for a rigorous mathematical foundation of Yang-Mills theory in physics, including a solution that explains why particles have mass.

Calculus

'Only geometry can hand us the thread that will lead us through the labyrinth of the continuum's composition, the maximum and the minimum, the infinitesimal and the infinite.'

GOTTFRIED LEIBNIZ

How can something be measured if it is continually changing – for example, the speed of an accelerating car? Efforts to answer this question reached their height with the arrival on the scene of 17th-century geniuses Isaac Newton and Gottfried Leibniz, inventing between them the modern concept of calculus. Calculus has revolutionized science more than any other area of mathematics, but its foundations require an understanding of mindboggling ideas, such as the infinitesimally small.

Calculus combines simple ideas in geometry with genius ideas about infinity, to create a way of measuring our changing world.

1 At any given moment in time nothing can be moving, because speed is distance divided by time and no time is passing.

TRUE / FALSE

2 The area of a circle can be found by cutting the circle into infinitely many wedges and rearranging these into a rectangle.

TRUE / FALSE

3 The symbol of integration, $\int$, comes from the letter 's' standing for 'summation'.

TRUE / FALSE

4 Differentiation (finding rates of change) and integration (finding areas) both use the ideas of infinitesimals, but are otherwise not related to each other.

TRUE / FALSE

5 It is impossible for a car to travel so that its acceleration is always equal to its speed.

TRUE / FALSE

TEN THINGS A GENIUS KNOWS

1 The difficulty in understanding motion
How is motion possible? The ancient Greek philosopher Zeno of Elea put his finger on the problem in his paradox of an arrow in flight. At any given moment in time, the arrow is not moving. In any photograph of an arrow it will be still. Yet if it does not move at any moment, then when does it move? A related problem comes when we try to calculate the speed of a car in a race. The formula for speed is distance divided by time, but the total distance of the race divided by the time taken will only provide an average speed. To find the speed of the car at a given moment, we will be trying to divide zero distance by zero time, which makes no sense.

2 How to use graphs to visualize rates of change
A function describes how one quantity changes with respect to another. For example, in a car race we can associate, to every moment of time, the distance that a car has travelled. Mathematicians use the notation $f(t)$, where t (in this example) is time and $f(t)$ is the distance travelled at time t. Often this is given by a formula such as $f(t) = t^2$, which would say that the distance is found by squaring the time. A graph of the function gives a visual image of the relationship between these variables: the time is plotted on the horizontal axis and the value of the function (that is, the distance) is plotted on the vertical axis. The speed of the car at a particular time is found by measuring the slope of the graph at that time. If the graph is curved, the slope is defined by drawing a tangent line: a straight line that just touches the curve at that point.

3 What differential calculus is
The rate of change of a function is called its derivative. The derivative is itself a function, describing how the slope of the tangent line changes over time. We can approximate the slope of the tangent line at a time $t = a$ by seeing how the distance has changed after a small time h and dividing this change in distance by the change in time. As a formula, we get the slope to be roughly $(f(a+h) - f(a))/h$. As h gets closer and closer to

zero, this fraction will get closer and closer to the true slope of the tangent line. The genius idea was to notice that, though the numerator and denominator are both becoming zero, the expression often still makes sense when h goes all the way to zero. For example, if $f(t) = t^2$ then:

$$\frac{f(a+h) - f(a)}{h} = \frac{(a+h)^2 - a^2}{h} = \frac{2ah + h^2}{h} = 2a+h,$$

which becomes $2a$ as h becomes zero.

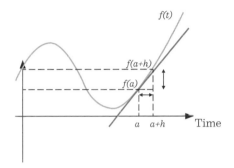

4 The difficulty in finding areas
Finding the area of a rectangle is easy: it is the width multiplied by the height. Finding the area of a triangle is also easy: it is half of the width multiplied by the height. The area of any shape defined by straight lines is then straightforward because it can be broken up into triangles. But how can the area of a curved shape be found? Archimedes struggled with this question when finding the area of a circle. His method was to cut the circle into wedges and say that the wedges were approximately triangles. By rearranging the triangles into a rectangle, he found a formula for the area of the circle. The problem was that to get an accurate answer, he needed to have infinitely many infinitely small wedges, and so had to multiply zero by infinity.

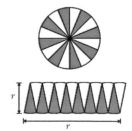

5 **What integral calculus is**
In the same way that differential calculus makes sense of zero divided by zero, integral calculus makes sense of infinity multiplied by zero. Integration of a function $f(t)$ means finding the area under the graph of $f(t)$. The genius idea is to break up the area into rectangles, each with a small width h. The height of each rectangle is taken to be the value of the function at one of the points within the base of the rectangle. The total area is then the sum of the areas of all the rectangles. As the value of h becomes smaller and the number of rectangles becomes bigger, this sum will become a better and better approximation for the area, and when h is zero it will be exact.

6 **Why calculus works**
Both differential and integral calculus are inherently problematic. They do calculations using a small (but positive) number h and then in the final step allow h to be zero. How can h be non-zero in the beginning and then zero at the end? The resolution to this paradox had to wait until the mid-19th century and the work on limits by Cauchy and Weierstrass (see page 141). The modern definition of integrals and derivatives involves taking the limit as h goes to zero, rather than simply setting h equal to zero at the end. This provides a rigorous framework for the subject and ensures the answers are correct.

7 **The notation used in calculus**
In both differential and integral calculus, mathematicians work with quantities that become infinitesimally small. The lowercase letter d is used to mean 'differential'. When written in front of a variable it means an infinitesimally small quantity of that variable. The notation for the derivative of $f(t)$ is:

$$\frac{df}{dt}$$

meaning an infinitesimally small change in the function f divided by an infinitesimally small unit of time. The area under the graph of $f(t)$ is denoted by $\int f(t)dt$, where the integral symbol $\int$ comes from the 's' of 'sum' because we are summing the areas of all the rectangles.

8 **How differentiation and integration are related**
A turning point in the development of calculus was the discovery of how integration and differentiation are related. It turns out that the two methods are inverses of one another. That is, if a function is integrated and then differentiated (or vice versa), it essentially remains unchanged. For example, if we differentiate a distance-time graph we get the velocity of a car. If we then integrate the velocity-time graph we get back the distance travelled. This result, called the fundamental theorem of calculus, was first fully developed as a theory by Isaac Newton, building on work by James Gregory and Isaac Barrow. Since integration is, in general, much more difficult than differentiation, the fundamental theorem of calculus gave scientists a practical way to perform integration using derivatives.

9 **The magic of the number e**
Differentiating or integrating a function produces a new function. For example, differentiating $f(t) = t^2$ gives $2t$; integrating $f(t) = \cos t$ gives $\sin t$. Is it possible to find a function that is unchanged by integration or differentiation? Such a function would describe its own rate of change, and it does indeed exist: it is $f(t) = e^t$, where e is a number with a rough value of 2.718. The letter e stands for 'exponential', and it is one of the foundational constants of mathematics, just as important and intriguing as π. Like π, the constant e is an irrational number, so its decimal expansion never repeats.

10 **Who invented calculus**
The big ideas of calculus were laid down over many centuries, from ancient Greek and Chinese scholars, to Indian, Middle Eastern and European mathematicians. Today, credit for the development of calculus goes to two men: England's Isaac Newton and Germany's Gottfried Leibniz. Although it is now widely accepted that Newton and Leibniz developed their ideas independently, accusations that Leibniz had plagiarized Newton's work caused a rift between British and continental mathematicians for over 100 years.

TALK LIKE A GENIUS

❝ The most famous critic of calculus was probably Bishop Berkeley, who, in 1734, wrote a treatise addressed to the "infidel mathematician". He described the infinitesimal increments as "ghosts of departed quantities", and objected to the way these were used as non-zero variables in the early part of a calculation but set to zero later on. In his opinion, calculus only worked because two different errors managed to cancel each other out. ❞

❝ The magic number e comes up in a question about compound interest. If I invest £1 and receive 100% interest after a year, my total is £2. But if I receive 50% interest twice a year, my total becomes £2.25, because I am collecting interest on the interest. Getting 25% interest four times a year brings my total up to £2.44. If interest were being paid continuously throughout the year, my total earnings would amount to £2.718, or e pounds. ❞

❝ In 2014, neuroscientist Semir Zeki and Fields Medallist Sir Michael Atiyah put mathematicians in an MRI scanner and studied their brains' responses to seeing mathematical formulae. The equations that the mathematicians had previously rated as being the most beautiful induced reactions in them that were the same as reactions to other forms of art like music or painting. And the formula that came out as being 'most beautiful' was Euler's identity relating the mathematical constants 0, 1, π and e: $1 + e^{i\pi} = 0$. ❞

WERE YOU A GENIUS?

❚ FALSE – Differentiation makes sense of how to define speed as the time variable approaches zero, so we can measure speed at each moment.

❷ TRUE – This method was known to the ancient Greek mathematician Archimedes and later made rigorous by integral calculus.

❸ TRUE – This notation was invented by the German mathematician Leibniz.

❹ FALSE – Differentiation and integration are the inverses of one another.

❺ FALSE – The exponential function $f(t) = e^t$ is its own derivative, so a car travelling at e^t m/s would be travelling at a speed that described its acceleration. (However, it would quickly exceed all speed limits!)

THE BLUFFER'S SUMMARY

Differentiation finds rates of change and integration finds the area under a curve. These two ideas together are called calculus.

To the limit

'A mathematician who is not somewhat of a poet will never be a perfect mathematician.'

KARL WEIERSTRASS

We all have an intuitive definition of a limit. A process or sequence that gets closer and closer to something and maybe only reaches it 'at infinity'. But intuition can be a dangerous thing in mathematics. Karl Weierstrass was the man who took the vague poetry of our intuition and turned it into unambiguous reality, earning the name the 'father of analysis'. The resulting framework may seem awkward and abstruse to outsiders – a mess of symbols only for geniuses to understand – but to mathematicians it has a poetry of its own. The epsilons and deltas of Weierstrass are the ABCs of the analyst, creating the language of calculus and forming the foundation of all that we take for granted in the modern world.

Mathematicians don't just push the limits of knowledge – they work to understand what limits are and how they matter.

ARE YOU A GENIUS ?

1 A sequence has a limit L if the elements of the sequence can eventually get arbitrarily close to L.

TRUE / FALSE

2 A divergent sequence is one that increases without bound.

TRUE / FALSE

3 Infinite decimals, like π, can be thought of as sequences of finite decimals.

TRUE / FALSE

4 If we take the expression $x/\sqrt{x}$ and look at what happens as x gets closer and closer to 0, the answer makes no sense because we cannot calculate 0 divided by 0.

TRUE / FALSE

5 It is possible to draw a continuous (unbroken line) on a sheet of paper that goes through every point on the paper and never crosses over itself.

TRUE / FALSE

TEN THINGS A GENIUS KNOWS

1 The intuition behind a limit

Think of a number. Halve it. Halve it again. Keep halving it to produce a sequence of numbers. For example, you might have chosen 8, 4, 2, 1, 0.5, 0.25, 0.125, ... The longer you continue this process, the closer you will get to 0, even though you will never hit 0 exactly. Mathematicians decided that they would call a number the limit of a sequence if the elements of the sequence eventually got arbitrarily close to the limit. In the halving example, the numbers in your sequence will eventually get smaller than 0.1. They will eventually get smaller than 0.00001. They will get closer to 0 than any (non-zero) distance we can think of.

2 Convergent versus divergent sequences

Not every sequence has a limit. Those that do, like the halving sequence, are called convergent, while those that do not are called divergent. Most sequences are divergent. Some sequences get bigger without bound – such as 1, 1, 2, 3, 5, 8, 13, ... Some sequences jump around and don't get closer to any particular number, such as 3, 7, 3, 7, 3, 7, 3, 7, ... Some might seem to be getting closer and closer to a number, but have 'outliers' that disrupt the convergence, like 0.1, 0.01, 5, 0.001, 0.0001, 5, 0.00001, ... Sequences can generate an infinite variety of behaviour and only a very tiny proportion behave in a way that we can understand.

3 The modern definition of the limit

Mathematicians from Archimedes through to Newton and Euler all used an intuitive notion of a limit to successfully calculate the results they needed. But by the 19th century, people were questioning the validity of these calculations. It was not enough to assume that limits existed – they had to be *proven* to exist. And for that, there had to be an unambiguous definition of a limit. The modern definition came from Karl Weierstrass in the 1870s, and uses the symbol ε (the Greek epsilon), to stand for the 'error' in the distance to the limit. A sequence is denoted by $a(n)$, so $a(1)$ is the first term, $a(2)$ is the second term and so on. We say that L is the limit of $a(n)$ if, for any positive error ε, there is a point in the sequence $a(N)$ after which every element of the sequence is within a distance of ε from L.

4 How to use limits to define real numbers

Any decimal number can be thought of as the limit of a sequence of finite decimals. For example, π is the limit of the sequence 3, 3.1, 3.14, 3.141, 3.1415, 3.14159 and so on. In the late 19th and early 20th centuries, when mathematicians were trying to make the real numbers rigorous, this became one way of actually defining what they were. When we say that 0.999 . . . = 1, what we mean is that the sequence 0.9, 0.99, 0.999 . . . has a limit equal to 1. The real numbers have a very special feature called completeness, which means that every convergent sequence of rational numbers has a real number as its limit.

5 From sequences to functions

The sequence 1, $1/2$, $1/3$, $1/4$, . . . has the formula $a(n) = 1/n$. In this formula, a natural number, n, is inputted, and a real number, $1/n$, comes out. We could create a very similar function $f(x) = 1/x$ where a number x goes in and another number, $1/x$, comes out, only here x could be any decimal number. This function can be plotted on a graph, with the inputs along the horizontal axis and the outputs along the vertical axis. This function fills in the gaps between the numbers in the original sequence, and it extends the range of questions we can ask about limits. Instead of only being able to ask what happens as x goes to infinity, we could also ask what happens as x approaches 1, or 0, or 352.45.

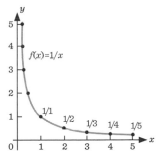

6 The limit of a function

The limit of a function has a slightly more complicated definition than that of a sequence, but the intuition behind it is the same. We are interested in what a function $f(x)$ does as x approaches a value c; for example, what happens to $1/x$ as x approaches 0. If a limit exists and is the number L, this is written as

$$\lim_{x \to c} f(x) = L.$$

To pass as a genius and impress your friends at dinner parties, you could do worse than to write on a napkin the definition of a limit. Mathematicians say that L is the limit of $f(x)$ as x approaches c, if we can make $f(x)$ arbitrarily close to L by making x move sufficiently close to c. In symbols, this comes out as: $\forall \varepsilon > 0 \, \exists \, \delta > 0$ such that if $|x - c| < \delta$ then $|f(x) - L| < \varepsilon$.

7 How to interpret the limit of a function
The symbols in the definition of a limit look incomprehensible to most people and even daunting to any mathematics student. It is perhaps illustrated by a metaphor. Suppose that we wish to bake a chocolate brownie. The variable x is the amount of flour and the function $f(x)$ is how good the brownie tastes if it has x amount of flour in it. We believe that as the amount of flour approaches 100g, the flavour gets closer and closer to its best flavour L. This can be tested as follows: for any given error ε we must be able to find a set of scales that can measure 100g to within a certain degree of precision δ, which will bake a brownie to within ε of the best flavour L. Clearly, if we want a brownie to be within 1% of the optimal flavour, we will need more precise scales than a brownie that is within 10% of the best flavour. The point is that, however good we want the brownie to be, there must be a set of scales that will allow us to achieve it.

8 Some surprising limits
Sometimes limits can give unexpected answers, which reinforces how important it is to rely on rigorous definitions and not on our intuition. For example, take the function $f(x) = \sqrt{x^2 - x} - x$. We can write $\sqrt{x^2 - x}$ as $\sqrt{x(x - 1)}$, and when x gets large, the numbers x and $x - 1$ are pretty much the same, like 999,999 and one million. So square rooting should give a number that is approximately equal to x. The function $f(x)$, then, should intuitively approach 0 as x goes to infinity – and yet, the limit is actually $-\tfrac{1}{2}$! Another notable example is that

$$\lim_{x \to \infty} (1 + \tfrac{1}{x})^x = e = 2.718\ldots,$$

which means there is a limit to how much we can earn by getting compound interest each year.

9 The idea of continuity
For the function $f(x) = \tfrac{1}{x}$ that we saw before, many of the limits are easy to find. The limit as x approaches 2 should be $\tfrac{1}{2}$, the limit as x approaches 3 should be $\tfrac{1}{3}$ and so on. This is because the function is continuous: its graph can be drawn in an unbroken line without any jumps or sudden changes. When a function is continuous, then we can find

$$\lim_{x \to c} f(x)$$

by simply putting in $x = c$ and calculating $f(c)$. In fact, this idea became the definition of continuity after Weierstrass's work, and is the foundation of the study of calculus, as functions can only be differentiated if they are continuous.

10 The relationship between fractals, continuity and differentiation
Every differentiable function must be continuous, but continuous functions are not always differentiable. Nowhere is this better demonstrated than in examples of fractals. Weierstrass himself was the first to find a function that was continuous everywhere but differentiable nowhere. A more famous example is in David Hilbert's space-filling curve (below). The curve is generated iteratively, each straight line in one iteration being replaced with a 'wiggly' line in the next iteration. Iterated to infinity, the resulting curve is continuous and touches every point in space, but changes direction all the time, making it impossible to do calculus on.

TALK LIKE A GENIUS

❝ Weierstrass's father sent him to university to prepare for a government position in finance, but Weierstrass secretly wanted to study mathematics. Torn between duty and inclination, he instead threw himself into drinking and fencing, leaving university without any degree at all. Later he taught himself mathematics and became so renowned that the University of Königsberg conferred an honorary degree upon him. ❞

❝ On the internet there sporadically appears a supposed "proof" that π is equal to 4. This starts by drawing a square around a circle with diameter 1, which makes the perimeter of the square equal to 4. It then proceeds by removing corners of the square, creating a stepped curve that becomes a better and better approximation to the shape of the circle. Despite this supposed approximation, the perimeter remains equal to 4 – so why is this not also the perimeter of the circle? The answer lies in the subtlety of limits. The area converges to that of the circle, but the perimeter does not converge to the circle's curved shape. ❞

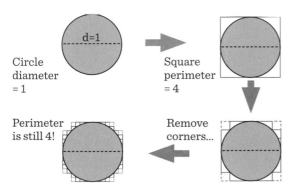

Circle diameter = 1

Square perimeter = 4

Remove corners...

Perimeter is still 4!

WERE YOU A GENIUS?

1 TRUE – For any stated distance, we must be able to find a point in the sequence past which all elements of the sequence are within that distance of the limit L.

2 FALSE – A divergent sequence is any sequence without a limit; such sequences may be bounded, such as 3, 7, 3, 7, 3, 7, ...

3 TRUE – Every real number can be thought of as the limit of a sequence of finite decimals, so π is the limit of the sequence 3.1, 3.14, 3.141, 3.1415, ...

4 FALSE – The expression $x/\sqrt{x}$ can be written more simply as $\sqrt{x}$, so as x gets closer to 0, this will also get closer to 0.

5 TRUE – Such a curve is called a space-filling curve, and has the strange property that it is changing direction at every point.

THE BLUFFER'S SUMMARY

If the outputs of a function get closer and closer to a number L as the inputs get closer and closer to a value p, then L is called the limit at p. This concept is the foundation of calculus.

Adding up infinity

'There you stand, lost in the infinite series of the sea, with nothing ruffled but the waves.'

HERMAN MELVILLE, *MOBY DICK*

Ancient philosophers, such as Zeno of Elea, through to present-day mathematicians and physicists have struggled with the idea of adding up infinitely many things and making sense of the answer. Modern analytical methods, together with the notion of convergence, have placed the idea of infinite sums on a rigorous footing, but there are still plenty of paradoxes and unanswered questions left to keep genius minds going.

Infinite sums can resolve ancient paradoxes, but they also create new ones.

1 Adding up infinitely many numbers always gives an answer of infinity.

TRUE / FALSE

2 Some of Zeno's paradoxes can be resolved using the mathematics of infinite sums.

TRUE / FALSE

3 The sequence $1 - 1 + 1 - 1 + 1 - 1 + ...$ has a value of 0.

TRUE / FALSE

4 If the terms in an infinite sequence get smaller and smaller, the sum of this sequence will be finite.

TRUE / FALSE

5 In some series, changing the order in which we add up the terms changes the value of the summation.

TRUE / FALSE

TEN THINGS A GENIUS KNOWS

1 What Zeno's paradoxes are

Ancient Greek philosopher Zeno of Elea used the concept of infinity to construct a set of ingenious paradoxes, which showed that change and motion were impossible. In the most famous one, Achilles and the tortoise are running a race. Since Achilles can run ten times faster, he allows the tortoise a 100-metre head start. By the time Achilles reaches the point where the tortoise started, the tortoise has moved 10m. By the time Achilles runs the next 10m, the tortoise has advanced a further metre. Each time Achilles catches up with where the tortoise used to be, the tortoise has moved forward, so Achilles can never catch it. In another paradox, Zeno wishes to walk across a room. But before he can do this, he must get halfway across, and before that he must cross a quarter of the way, then an eighth, a sixteenth and so on, meaning that he must do infinitely many things before he can begin to move.

2 How to rephrase Zeno's paradoxes as infinite sums

In Zeno's first paradox, we can calculate the distance that Achilles must travel by adding up an infinite sequence: 100m + 10m + 1m + 0.1m + 0.01m + ... In the second paradox we are asked to add up $1/2 + 1/4 + 1/8 + 1/16$ ..., with the distances halving every time. Mathematicians asked whether it was possible for these infinite sums (called series) to have a finite answer, thereby resolving the paradoxes. And if they do, how can the answer be calculated? Infinite sums are commonly written using the symbol Σ (sigma) to mean 'sum'. For example, the sum in the second paradox, where the denominators are increasing powers of 2, would be written more succinctly as

$$\sum_{n=1}^{\infty} 1/2^n$$

which means adding up the terms $1/2^n$ as n runs from 1 to infinity.

3 How to use algebra to find the value of an infinite sum

The sequence $1/2$, $1/4$, $1/8$, $1/16$, ... can be visualized as successive areas inside a square of side length 1. It is clear that as we add more of these terms together, they will fill out the whole area of the square,

indicating that the infinite sum will be equal to 1, the area of the square. This can be proved by doing some algebra. Let S stand for the series $1/2 + 1/4 + 1/8 + 1/16$, ... Then $S/2$ is the sum where all the terms are halved, giving $1/4 + 1/8 + 1/16 + 1/32$, ... If we calculate the difference between these sums, $S - S/2$, we see that all the terms cancel out except for the initial $1/2$. So $S - S/2 = 1/2$, meaning $S/2 = 1/2$, meaning $S = 1$, confirming our intuition from the square.

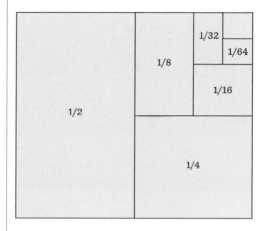

4 What a divergent series is

Not all infinite sequences have a finite sum. Sequences that do not are called divergent series. There are some obvious examples, such as $1 + 1 + 1 + 1 + ...$, and, in general, any sequence where the terms are not decreasing. There are also some paradoxical examples, such as Grandi's series: $1 - 1 + 1 - 1 + 1 - 1 + ...$. By summing the terms in different orders we can get two possible answers. The series $(1 - 1) + (1 - 1) + (1 - 1) + ...$ gives a sum of 0, but $1 + (-1 + 1) + (-1 + 1) + ...$ gives a sum of 1. Algebraic methods can be used to give a value of $1/2$, but most mathematicians accept that this kind of infinite sum does not have a sensible answer.

5 How to tell if a sequence has a finite sum

The algebraic method for summing infinite series is not rigorous by modern mathematical standards, as is seen by the problems with Grandi's series. Mathematicians therefore needed to agree on a method of deciding whether a series was finite. They did this by looking at the sequence of partial sums; that is, the sequence of numbers made by taking the sum of the first two terms, then the first three terms and so on. If this sequence of numbers converges to something (see page 141), then that number is called the sum of the sequence. Grandi's series has partial sums of 0, 1, 0, 1, 0, . . ., which do not converge, showing that the series is divergent.

6 The importance of geometric series
In a geometric series, each term is obtained from the previous one by multiplying by a constant factor, called the common ratio. In Zeno's paradox of Achilles and the tortoise, the common ratio is $1/10$ (the next distance Achilles must cover is ten times smaller than the previous one), while in the paradox of the room the common ratio is $1/2$. If the first term in such a series is denoted by a and the common ratio by r, then the series will converge (give a finite answer) whenever r is between -1 and 1, and the value of the series will be $a/(1-r)$. So we can calculate the distance when Achilles will overtake the tortoise by substituting $a = 100$ and $r = 1/10$, giving an answer of $100/0.9$, which is roughly 111.11m.

7 Whether the harmonic series diverges
If the terms of a sequence of numbers are not getting smaller, then their sum will clearly be infinite. But the converse is not true: just because the terms of a sequence are getting smaller, it does not mean that the infinite sum will converge. The best example of this is the harmonic series, which is the reciprocal sums of the natural numbers: $1 + 1/2 + 1/3 + 1/4 + 1/5 + \ldots$. Although the terms are getting smaller, the partial sums increase without bound, showing that this is a divergent series. One proof of this is as follows. The harmonic series is bigger than the series $1 + (1/2) + (1/4 + 1/4) + (1/8 + 1/8 + 1/8 + 1/8) + (1/16 + \ldots + 1/16) + \ldots$ (this is easily seen by comparing the two series term by term), and this second series is equal to $1 + 1/2 + 1/2 + 1/2 + \ldots$, which is clearly infinite.

8 How to make an alternating series
It is possible to have an infinite sum where some of the terms are negative. If the terms alternate between being negative and positive, this is called an alternating series. The alternating series $1/2 - 1/4 + 1/8 - 1/16 + \ldots$ is equal to $1/3$, while the more interesting

alternating series $1 - 1/3 + 1/5 - 1/7 + \ldots$, with the denominators running through the odd numbers, gives the surprising value of $\pi/4$. Although the harmonic series does not converge, the alternating harmonic series does: $1 - 1/2 + 1/3 - 1/4 + \ldots$ has a value of $log(2)$, the natural logarithm of 2.

9 Why the Riemann series theorem is so counterintuitive
The alternating harmonic series is an example of a series that is conditionally convergent. This means that it does converge, but if we make all the negative terms positive then it no longer converges. Such a series has a remarkable property, discovered by the German mathematician Bernhard Riemann and encapsulated in the Riemann series theorem. It says that any conditionally convergent series can have its terms rearranged so that the sum gives any value we want. So the alternating harmonic series $1 - 1/2 + 1/3 - 1/4 + \ldots$ has a value of $log(2)$, but by adding the terms in a different order the sum could be -1, or π, or even infinity.

10 Sequences whose sums are unknown
There are some sequences where mathematicians have been unable to decide yet whether their sum converges to a finite value. Such open problems usually point to a lack of understanding of a deeper area of mathematics. One example is the Flint Hills series

$$\sum_{n=1}^{\infty} \frac{1}{n^3 (\sin n)^2}$$

whose convergence or divergence will depend on how well the number π can be approximated by rational numbers (integer fractions). Another example relies on our understanding of prime numbers: the series

$$\sum_{n=1}^{\infty} (-1)^n \frac{n}{p_n}$$

where p_n is the nth prime number. Interestingly, we do already know that the sum of the reciprocals of the primes converges, as well as the reciprocals of the twin primes (pairs of primes differing by 2).

❝ The divergence of the harmonic series provides some very counterintuitive results. One of these is that it is possible to stack a set of identical blocks on the edge of a table with any possible overhang. One block overhangs the next by $1/2$, then that one overhangs the next by $1/3$, which overhangs the next by $1/4$ and so on. Since the harmonic series diverges, the total overhang can be as large as we want it to be. ❞

❝ The harmonic series only just diverges. If we remove all the terms whose denominator contains a 9, then the resulting series converges. In fact, if we remove all the terms whose denominator contains any finite sequence of digits, for example 1429763, then the resulting series will converge. ❞

❝ The reciprocals of the square numbers $1, 1/4, 1/9, 1/16 \ldots$ add up to the surprising value of $\pi^2/6$. Taking the sum of reciprocals of fourth powers, or sixth powers, or any even power always gives an answer that can be expressed in terms of π, but it is unknown whether a similar formula exists for the reciprocals of odd powers. It is not even known whether these odd power summations always give an answer that is an irrational number. ❞

1 FALSE – Some infinite sums have a finite answer.

2 TRUE – The paradoxes of Achilles and the tortoise, and the impossibility of crossing a room, can both be resolved using the methods of geometric series.

3 FALSE – This sequence does not have a limit, as the partial sums are 1, 0, 1, 0, 1, . . . and this sequence of numbers does not converge.

4 FALSE – A counterexample is the harmonic series, which diverges even though each term is smaller than the last.

5 TRUE – This happens in conditionally convergent series as a result of the Riemann series theorem.

THE BLUFFER'S SUMMARY

Sometimes it is possible to sum an infinite number of things and get a finite answer.

Modelling change

'The only constant is change, and the rate of change is increasing.'

PETER DIAMANDIS

Differential equations are the mathematical tools that help us understand our continually changing world. Whether it is the flight of a cricket ball, the spread of flu, the movement of the planets around the Sun or the vibrations of a guitar string, differential equations are behind every aspect of our lives. Sometimes we have the tools to solve these equations, but most of the time they are too difficult and the best we can do is use computers to find numerical approximations to the solutions. It is a topic sure to keep mathematical geniuses busy for many centuries to come.

Many natural phenomena change in highly complex and often self-referential ways – differential equations are how mathematicians understand what is going on.

ARE YOU A GENIUS

1 If the rate of change of a population depends on the size of the population, then this population will grow linearly over time.
TRUE / FALSE

2 The rate of change of the rate of change of distance with respect to time is more commonly known as acceleration.
TRUE / FALSE

3 If each rabbit has, on average, three babies each month and we start with two rabbits, then after a year there will be over eight million rabbits.
TRUE / FALSE

4 The spread of a disease gets faster and faster as it moves through a population.
TRUE / FALSE

5 A numerical method is one that finds an approximate solution, not an exact solution.
TRUE / FALSE

TEN THINGS A GENIUS KNOWS

1 **The idea behind a differential equation**
Differential equations describe how rates of change are themselves changing. A hiker on a walk may find that their hunger levels get steadily higher the further they walk. In this situation, the rate of change of hunger with respect to distance is a constant. A plot of hunger against distance would show a straight-line graph. In contrast, consider the rate of change of a population of rabbits. If each rabbit, on average, has three babies, then starting with two rabbits gives $(2 + (2 \times 3)) = 8$ rabbits after one generation, then $(8 + (8 \times 3)) = 32$ rabbits after two generations, and $(32 + (32 \times 3)) = 128$ rabbits after only three generations. Here the growth rate of the population is proportional to the size of the existing population, and the behaviour is very different from the changing hunger of the hiker.

2 **What a differential equation looks like**
Any differential equation starts with a function describing the relationship between two variables. For example, $h(x)$ might be the hunger levels of the hiker after they have walked a distance of x miles, or $R(t)$ may be the number of rabbits after a time t. The rate of change of one variable with respect to another is written as the differential, for example $\frac{dh}{dx}$ or $\frac{dR}{dt}$. A differential equation is then a formula involving this rate of change. For the hiker, it would be $\frac{dh}{dx} = c$, where c is a constant number. For the rabbits, it would be $\frac{dR}{dt} = rR$, where r is a number encapsulating the relative growth rate, taking into account both births and deaths.

3 **The language of differential equations**
Some differential equations involve the rate of change of the rate of change, and possibly higher order rates of change. To calculate the position of a skydiver $x(t)$ at a time t, we must solve a differential equation involving not only their speed, $\frac{dx}{dt}$, but also their acceleration

$$\frac{d^2x}{dt^2} = \frac{d}{dt}\left(\frac{dx}{dt}\right)$$

due to gravity and air resistance. The resulting equation is called a second-order differential equation because the highest derivative here is the second derivative. When talking about differential equations, it is common to state, along with the order, whether they are linear or non-linear. In a linear equation the function and its derivatives never appear as a square or a cube or any power, and neither is the function allowed to appear in a product with any of its derivatives.

4 **How to model with differential equations**
Differential equations are used to model real-world situations, and there is always a choice to be made of the level of detail put into a model. Simple equations may be easy to solve, but they may not give very accurate answers. Adding in more detail may make the equation more realistic, but at the expense of making it hard to solve. The rabbit population model is one example of this. The differential equation in section 2 is a simple model of how the rabbit population grows, and is easily solved via integration to show that the population grows exponentially over time. A more realistic model would be

$$\frac{dR}{dt} = rR\left(1 - \frac{R}{N}\right),$$

where N is the maximum population that the environment can hold. This is called the logistic equation, and is a non-linear equation because on the right-hand side there is an R^2 when the brackets are multiplied out. In this case we do know how to solve the equation but, in general, non-linear equations are very difficult to deal with.

5 **How differential equations may be coupled**
Suppose our rabbits live in an area where there is also a population of foxes. The more foxes there are, the smaller the rabbit population will become. A decreasing rabbit population will cause a decrease in the fox population because there isn't enough food. But a decrease in the number of foxes makes the rabbit population increase again, and so the cycle continues. In this situation, the rate of change of the rabbits depends on the number of foxes, and the rate of change of the foxes depends on the number of rabbits, resulting in two coupled differential equations. These are called the Lotka-Volterra equations and are widely used to describe predator-prey dynamics.

6 **What the SIR model is in epidemiology**
Modelling the spread of a disease is also done using coupled differential equations. It breaks the population into three groups: those who are susceptible to the disease, S, those who are infectious, I, and those who have recovered and are then resistant, R. The value of S will decrease over time as people contract the disease, so $\frac{dS}{dt}$ is inversely proportional to I. The value of I grows in proportion to both S and I because the disease grows fastest with lots of infectious people and lots of susceptible people. But it will also decrease as people recover from the disease, a factor proportional to R. Finally, the value of R grows in proportion to the number of infected people I. These three coupled differential equations are called the SIR model, and it is used by governments to decide when a disease is dangerous enough to require interventions such as vaccination or quarantine.

7 **What a numerical method is**
Sometimes a differential equation, or a system of coupled differential equations, cannot be solved analytically. This means that they cannot be solved using pen and paper, manipulating the symbols according to algebra and calculus, to get an exact formula for the solution. In these cases, mathematicians look for a numerical solution, which means an approximate solution. Numerical methods use very clever, mathematically justified, systems of guesswork in order to gradually hone in on the answers. Different methods are used for different types of differential equations, and there is a whole subject, called 'numerical analysis' that investigates how quickly these methods work and how accurate they are.

8 **How Euler's method works**
Euler's method is the earliest known numerical method for solving differential equations, and is still the basis of many algorithms used today. The idea is that we want to find the shape of a curve, and all that we know is its starting point and a differential equation that tells us its slope. Input the starting point, labelled A_0, into the differential equation to get a value for the slope at that point. Take a small step of size h along this line to get to a new point A_1. This new point may not be exactly on the curve we seek,

but if h is small then it will be very close. Repeat the procedure to generate a whole sequence of points, that will then form an approximation to the curve we were looking for.

9 **How good Euler's method is**
The smaller the step size h in Euler's method, the better the approximate curve will be. It turns out that the global error is proportional to h, so if we halve the step size, then our error will halve too. For many applications this error is impractical, requiring extremely small step sizes to get a good solution, which in turn means a very large number of steps. A better suite of numerical methods, called Runge-Kutta methods, make use of a set of weighted midpoints between the A_i to improve the approximation without increasing the number of steps needed. Numerical methods like these are used extensively in physics, engineering, biology, chemistry and economics.

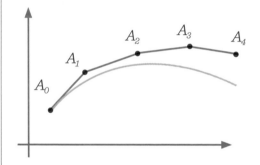

10 **The difference between an ODE and a PDE**
Differential equations come in two flavours: ordinary differential equations (ODEs) and partial differential equations (PDEs). ODEs involve functions of a single variable, as in all the examples we have seen, while PDEs involve functions of more than one variable. Modelling the vibration of a guitar string requires keeping track of the displacement at each point along the string over time, which requires a second order PDE in two dimensions. Unsurprisingly, PDEs are generally much harder to solve than ODEs, and non-linear PDEs in particular are considered some of the most difficult mathematical problems that we have.

TALK LIKE A GENIUS

❧ In the film *Hidden Figures*, about the lives of three African–American women working at NASA, a group of engineers are stumped with the problem of how to bring astronaut John Glenn's capsule safely back to Earth. Having tried in vain to solve the differential equations using geometry, mathematician Katherine Johnson saves the day by suggesting they try Euler's method to get the solution. ❧

❧ Einstein's field equations for general relativity are a set of ten coupled non-linear partial differential equations. Exact solutions are only known in special situations with a high degree of symmetry. This is how black holes were found – when Karl Schwarzschild found a solution by assuming that all mass was concentrated at a single point. Other solutions have been found that predict phenomena like wormholes, but there is no guarantee that these will actually exist in our universe. ❧

❧ The Black-Scholes equation is a partial differential equation that is widely used in modelling financial markets. It estimates the prices of options over time and allows bankers to buy and sell assets at just the right time to make profit without risk. Some say that the equation contributed to the 2007 banking crisis, but others defend it, saying that its users should have been more aware of its limitations and assumptions. ❧

WERE YOU A GENIUS?

1 FALSE – Such a population may, for example, grow or decay exponentially, but it will not change linearly.

2 TRUE – The rate of change of distance with respect to time is speed, and the rate of change of speed is the acceleration.

3 TRUE – Such a situation happened in Australia, where rabbits had no natural predators. Just 24 rabbits released in 1859 had swelled to ten billion by the 1920s.

4 FALSE – This may be true in the initial stages of an outbreak, but as more people recover and become immune, the spread of the disease will slow down.

5 TRUE – Not all differential equations can be solved exactly, and in these cases numerical methods are the best technique we have of finding solutions.

THE BLUFFER'S SUMMARY

Differential equations describe how rates of change are themselves changing and are a key tool in modelling many areas of science, engineering and economics.

The Navier-Stokes equations

'When I meet God, I am going to ask him two questions: Why relativity? And why turbulence? I really believe he will have an answer for the first.'

WERNER HEISENBERG
(ALSO ATTRIBUTED TO HORACE LAMB)

The Navier-Stokes equations explain how the oceans move around the Earth, how blood moves around our bodies, and how air moves over the wing of an aeroplane. Despite their usefulness, we still have no idea how to solve them, though it has been over 150 years since they were written down. Such is the importance of finding a solution that the problem has been named as one of the Millennium Prize Problems.

Turbulence has been called the biggest unsolved problem in physics – a solution to the Navier-Stokes equations would offer at least a starting point.

ARE YOU A GENIUS
?

1 A fluid is a liquid, such as water, oil or milk.
TRUE / FALSE

2 The viscosity of a fluid is a measure of the amount of friction in the fluid.
TRUE / FALSE

3 The faster a fluid is flowing, the lower the pressure it is exerting.
TRUE / FALSE

4 It is possible to win a million dollars by solving the Navier-Stokes equations in the simplest case where all the external forces are zero.
TRUE / FALSE

5 Turbulence is when a fluid suddenly starts flowing quickly.
TRUE / FALSE

TEN THINGS A GENIUS KNOWS

① What a fluid is

The Navier-Stokes equations describe the motion of a flowing fluid. A fluid need not be a liquid, such as water or oil, but could also be a gas, plasma or a type of solid. The definition of a fluid is any substance that cannot resist a shear force applied to it, or alternatively any substance that flows to take the shape of the container it is in. Although, in reality, a fluid is made up of lots of tiny particles, mathematicians model it as a continuous medium. For the Navier-Stokes equations there are further simplifying assumptions to make the problem tractable: the fluid is assumed to be homogenous (it looks the same everywhere) and incompressible (it has the same density everywhere).

② What things affect the flow of a fluid

The Navier-Stokes equations are based on Newton's second law, which says that the force applied to an object is equal to its mass multiplied by its acceleration, or $F = ma$. What are the forces acting on a flowing fluid? One of the forces is pressure. Divers will experience higher pressures the deeper they swim, due to the weight of the water above them. Similarly, we all experience atmospheric pressure from the weight of the atmosphere pressing down on us. There may be external forces, such as gravity, a centrifugal force or a magnetic force. Finally, there is viscosity, which is a measure of the frictional forces between moving layers of the fluid. Honey is more viscous than water, which is why it flows much more slowly.

③ How Bernoulli's equation explains how aeroplanes work

The first person to write down a formula for how fluids flowed was the Swiss mathematician Daniel Bernoulli, in 1738. One of the results his equation produced said that a fluid flowing over a surface would exert a smaller pressure the faster it travelled. This is exactly the principle used to explain how aeroplanes work, although it was to be more than a century until man-made wings were created. If air is moving faster over the top of the wing than underneath it, the pressure will be lower on the top than the bottom, and the net effect is upward lift of the plane. The next significant work on the study of fluids was made by Euler, who built on Bernoulli's work to describe a set of equations for the motion of a viscous-less fluid. To do this, he needed a new kind of calculus.

④ How calculus works in multiple dimensions

When a fluid is flowing, it is doing so in three-dimensional space and another dimension of time. The study of motion is calculus, but traditional calculus as developed by Newton and Leibniz was only concerned with how one variable changed with respect to another. In a river flowing down a hill, we want to capture the change in three variables at once: the vertical speed, the forward speed and the side-to-side speed. The tool for doing this kind of analysis was gradually developed in the century after Newton, and is called partial differentiation. It holds all but one of the variables fixed while applying standard calculus techniques along the last direction.

⑤ What the notation is for partial differentiation

A function is a formula that produces an answer for every input. The input can have any number of variables. For example, $T(x,y,z)$ might give the temperature on Earth at a latitude of x, a longitude of y and an altitude of z. Or $g(d,s)$ might give the remaining petrol in your tank after you have driven d miles at an average speed of s. To describe how temperature is changing with latitude, we would write $\frac{\partial T}{\partial x}$ and call this the partial derivative of T with respect to x. Similarly, the rate of change of petrol with respect to your distance would be $\frac{\partial g}{\partial d}$ and with respect to your average speed would be $\frac{\partial g}{\partial s}$.

⑥ Euler's equations for fluid flow

Euler's equations use four different functions to describe how a fluid evolves over the dimensions of space, (x,y,z), and time, t. These functions are: v_x, the velocity in the x-direction; v_y, the velocity in the y-direction; v_z, the velocity in the z-direction; and p, the pressure of the fluid. For example, $v_x(0.5, 1, 7, 60)$ would be the horizontal velocity of the fluid at the spatial coordinates $(0.5, 1, 7)$, 60 seconds after the flow had started. To model the fluid flow, Euler also took into account possible external forces f_x, f_y and f_z acting in the x-, y-, and z- directions, respectively. This created three equations of the form:

$$\frac{\partial v_\alpha}{\partial t} + v_x \frac{\partial v_\alpha}{\partial x} + v_y \frac{\partial v_\alpha}{\partial y} + v_z \frac{\partial v_\alpha}{\partial z} = f_\alpha - \frac{\partial p}{\partial x},$$

one for α being each of x, y and z. A final equation expressing the condition that the fluid was incompressible meant that there were four equations for the four unknown variables, which was important because, otherwise, there could be an infinity of solutions.

7 **How the Navier-Stokes equations differ from Euler's equations**
The Navier-Stokes equations are very similar to Euler's equations, except that they add in an extra term to account for the viscosity of the fluid. The viscosity is given the symbol ν (the Greek letter *nu*), and though it might seem like a small detail to add in, it greatly increased the complexity of the equations. This is because it involved adding in terms like

$$\nu \left(\frac{\partial^2 v_x}{\partial x^2} + \frac{\partial^2 v_x}{\partial y^2} + \frac{\partial^2 v_x}{\partial z^2} \right),$$

which use second partial derivatives. Here, for example, we differentiate with respect to x and then differentiate with respect to x again. It is worth noting that even Euler's equations have not been solved, and the Navier-Stokes equations are more complicated still.

8 **How to write the Navier-Stokes equation in terms of vectors**
The Navier-Stokes equations look very daunting when they are written out in their full form, but, using the ideas of vector calculus, they can be made a little more palatable. A vector is a mathematical object that captures both size and direction. The three equations describing fluid flow are all aspects of the same thing, just that each looks in a different direction. Writing **v** as the velocity vector and **f** as the vector of forces, we can rephrase the three separate Navier-Stokes equations as the single equation:

$$\frac{\partial \mathbf{v}}{\partial t} + (\mathbf{v} \cdot \nabla)\mathbf{v} = \mathbf{f} - \nabla p + \nu(\nabla \cdot \nabla)\mathbf{v},$$

where ∇ and $\nabla \cdot$ denote the vector calculus operations of gradient and divergence respectively.

9 **How to win one million dollars with the Navier-Stokes problem**
The Navier-Stokes equations were first written down by Frenchman Claude-Louis Navier in 1822, and derived rigorously by Irishman George Gabriel Stokes in the 1840s. Over 150 years later, we still have no idea how to solve them. The problem is so difficult that the rules of the Clay Mathematical Institute allow for its million-dollar Millennium Prize Problem to be won in several ways. In the simplest version, you can assume that all external forces are zero, and from there find a set of functions that satisfy the Navier-Stokes equation. (These solutions also need to be mathematically well behaved, or *smooth*.) Alternatively, you can win the money by finding a set of external forces for which the equations have no solution at all.

10 **Progress on the problem so far**
The Navier-Stokes equations are known to have well-behaved solutions when they are restricted to fluids moving in only two dimensions. They are also known to have solutions when the fluid starts out moving very slowly. In the general three-dimensional case, though, progress in solving the equations has been very slow. People have shown that each fluid has a 'blowup time', before which the equations can be solved, but after which infinities may creep in and create singularities. The difficulty in making progress can be linked to the phenomenon of turbulence, which is the appearance of chaotic behaviour in fluids and which is going to require extremely complex equations to describe.

TALK LIKE A GENIUS

❝ It is a common myth that glass is a liquid. This myth is perpetuated by people who notice that in old buildings such as churches, the glass is often thicker at the bottom of a pane than at the top. This seems to indicate that the glass has flowed over time. While glass does flow in its liquid state, it does not continue to do so once it has solidified. For the glass in old windows to have achieved its shape because of flow, it would need longer than the age of the universe to have done so. ❞

❝ From Bernoulli's equation, we know that a wing generates lift when air flows faster over the top than underneath it. Many people erroneously believe that this happens because the shape of the wing forces air to travel a longer distance over the top than underneath, and so travel faster. This cannot be true, because it is possible for planes to fly upside down! Lift is actually created by tilting the wing at an angle to the air stream, forcing the air downwards. ❞

1 FALSE – A fluid is any substance that takes the shape of the container it is in, so can include liquids, gases, plasmas and even some solids.

2 TRUE – This means that more viscous fluids flow slower than less viscous fluids.

3 TRUE – This explains how planes take off, as air flows faster over the top of a wing than below, lowering the air pressure on top and creating upward lift.

4 TRUE – The problem is considered so difficult that the prize money will be awarded even if this simple case is solved.

5 FALSE – Turbulence manifests as a fluid undergoing rapid fluctuations in speed and direction, and is when the flow becomes mathematically chaotic.

THE BLUFFER'S SUMMARY

The Navier-Stokes equations model the flow of a viscous fluid. One million dollars can be won by anyone who either solves them, or shows that they cannot be solved.

Chaos theory

'The things that really change the world, according to chaos theory, are the tiny things.'

NEIL GAIMAN

Chaos, contrary to popular belief, does not mean randomness. Mathematicians use the word to describe systems where a tiny initial change (such as a butterfly flapping its wings in the Amazonian jungle) results in a wildly different outcome (such as a storm in Europe). Unpredictability arises not because we do not understand the equations governing the system, but because measurements are never perfect. It explains why our weather forecasts will never be very good and why people came to believe that lemmings commit mass suicide.

The discovery of chaotic systems put limits on scientific prediction, but also revealed complex and beautiful patterns that scientists are still trying to understand.

ARE YOU A GENIUS

1 Chaotic systems are deterministic, so the future is exactly determined from the present.

TRUE / FALSE

2 Every real world system is chaotic to some degree.

TRUE / FALSE

3 A pendulum swinging on the bottom of another pendulum traces out a similar pattern to a regular pendulum.

TRUE / FALSE

4 Lemming populations will occasionally almost die out because they have too many babies.

TRUE / FALSE

5 If computers were twice as powerful, we could predict the weather twice as far into the future.

TRUE / FALSE

TEN THINGS A GENIUS KNOWS

① The difference between deterministic and stochastic systems

A dynamical system is a way of modelling the evolution of something in time. The something might be the angle of a pendulum, or the number of sheep in a field, or the location of a drunken person attempting to walk home. Dynamical systems come in two varieties: deterministic and stochastic. A stochastic system, like the drunken walk, has an element of randomness, or probability, to the equations, while a deterministic system does not. In a deterministic system, the equations governing the evolution of the system will tell you exactly what will happen to any point as it moves through time.

② What happened to Edward Lorenz

In 1961, the American mathematician and meteorologist Edward Lorenz was running a computer simulation of a weather system. The equations used 12 different variables, representing things like temperature, pressure and wind speed, and were so complex that the calculations of the evolution of the system could not be done by hand. Even with a computer they took a long time. So one day, in order to get further with a simulation, Lorenz restarted a computation from the middle instead of the beginning, using as input the computer readouts of the day before. Returning after a coffee break, he found that this simulation produced completely different results from the day before, even though it should have been using the same data. It was the beginning of the modern study of chaos.

③ What chaos is

In a chaotic system, a very small change to the initial conditions will result in a wildly different outcome. This is very different from our intuition and everyday experience. If we make a tiny miscalculation about the right place to kick a football, we still expect the football to get very close to where we wanted it to go. If we start two pendulums swinging at angles 0.01° different from one another, we expect their trajectories to be almost identical. What Lorenz had found was a system where a tiny rounding error – he had used a computer readout of 0.506 when the true value was 0.506127 – resulted in a completely different prediction from before.

④ The two ingredients of chaos

Sensitivity to initial conditions is not the only detail that defines a chaotic system. It is easy to find equations in which points that are close together in the beginning end up being far apart, but that we do not think of as chaos. One example is the doubling map, which takes a number and keeps on doubling it. It stretches out numbers on the number line, so will make any initial pair of points move far apart, but there are no surprises in the way the numbers behave. A chaotic system must also demonstrate the phenomenon of 'mixing', where the trajectories of any two collections of points must eventually overlap with each other. In general, a combination of stretching and folding, like a baker kneading dough, is what produces a chaotic system.

⑤ How pendulums can be chaotic

The standard swinging pendulum is not chaotic, but simple changes can introduce chaos into the system. One way is to attach a second pendulum onto the bottom of a first one, creating what is known as a double pendulum (below left). Tiny changes in the initial configuration of the pendulum will result in very different trajectories, with each trajectory itself being highly complex, as traced in the timelapse photo below right. The equations governing the pendulum are known – any unpredictability is solely from errors in measuring the initial position. A second way to make a chaotic pendulum is to hang an iron ball on a piece of string, then place a number of magnets beneath it, just out of reach. The pendulum's path is perturbed by the magnets, resulting in chaotic motion.

⑥ What the logistic map is

The logistic map is a simple equation used to model animal populations, demonstrating a surprising range of complex and chaotic behaviour, despite its simplicity. The idea is that two things influence the growth of a collection of animals: the

birth rate and the maximum population that the environment can hold. The population should grow when there are small numbers of animals, but should shrink if the population is reaching its maximum size, due to overcrowding and competition for food. This can be captured in the equation $x_{n+1} = r\,x_n\,(1 - x_n)$, where r is 1 plus the growth rate of the population, and x_n is the size of the population at time n, adjusted so that the maximum population has size 1.

7 The behaviour of the logistic map

Different values for r, the reproductive variable, wildly change the behaviour of an animal population. If r is less than 1, the population will die out because they're not reproducing enough. When r is between 1 and 3 the population reaches a fixed size, though it may fluctuate around that value. With r between 3 and 3.45 the population size switches between two different values, then up to $r = 3.54$ it fluctuates between four values, then eight, then sixteen. So far, so good. But when r hits a magic value of approximately 3.57 the logistic map suddenly predicts chaos. The population size varies erratically, and tiny changes in any variable give a very different graph of population growth. Scientists believe that this chaotic behaviour may help explain why lemming populations exhibit unpredictable population booms and busts – contrary to the popular belief that they commit mass suicide.

8 What the bifurcation map is

In the logistic map, we see an example of 'period doubling' – the number of steady states of a population are first 1, then 2, then 4, 8, 16, …. This can be shown graphically on a bifurcation diagram, which shows the values of r at which the behaviour of the system changes. At $r = 3.57$ we see the onset of chaos. More interestingly, as the diagram continues, we see that it has a fractal structure: zooming in on portions of the diagram reveal the same structure as the whole picture. This also reveals 'islands of stability' – fleeting values of r at which the population of animals once again becomes stable and varies predictably between 1, 2, 4 or 8 values. The connection between fractals and chaos is not unusual and appears in many different systems.

9 How to spot a strange attractor

An attractor of a system is a set of points towards which trajectories of the system evolve. In the logistic map when $r = 3$, the population size gets closer and closer to a fixed value, so this value is called an attractor. In more complicated systems the attractor may be a curve or a surface, rather than just a single point. Chaotic systems often have attractors with a fractal structure, and these are called strange attractors. The most famous example is probably that of the Lorenz attractor, which resembles the wings of a butterfly and is a simplified model for atmospheric convection.

10 The consequences of chaos

When a system is chaotic, it becomes almost impossible to make long-term predictions because no measurement is ever completely accurate. For example, even if we understood the exact equations and factors that govern the motions of our atmosphere, and even if computing power increased and measurement instruments were to become very accurate, we would still not be able to predict the weather very far in advance, because the equations are chaotic. Other examples of chaotic systems include planetary motion, stock-market fluctuations, turbulence, fluid mixing and chemical reactions.

TALK LIKE A GENIUS

❧ Chaos theory is often referred to as the "butterfly effect", though it is unclear from where this phrase originates. Lorenz himself may have coined it by giving a talk in 1972 titled "Does the flap of a butterfly's wings in Brazil set off a tornado in Texas?", though a history-changing butterfly was also the plot of a sci-fi novel by Ray Bradbury in 1952. In *A Sound of Thunder* a time traveller accidentally crushes a butterfly in the time of the dinosaurs, which then changes the future. ❧

❧ Climate predictions and weather predictions are different. While we cannot say whether it will be snowing on Christmas Day 2020, we can predict with confidence what the average global temperature will be that year. Climate is not chaotic in the way that weather is. ❧

❧ Four out of Pluto's five moons rotate chaotically. Not only would every day be a different length on these moons, but there would be no certainty on where the Sun would rise or set each day. According to astronomer Mark Showalter, "The moon Nix can flip its entire pole. It could actually be possible to spend a day on Nix in which the Sun rises in the east and sets in the north." ❧

WERE YOU A GENIUS?

1 TRUE – Chaotic systems are governed by exact equations without any randomness in them, so each input gives a fixed output.

2 FALSE – Only certain systems exhibit true chaos, showing both sensitivity to initial conditions and 'mixing'. For example, a standard pendulum swing is not chaotic.

3 FALSE – Double pendulums exhibit chaotic motion, tracing out complex trajectories that vary wildly depending on small changes to the initial angle.

4 TRUE – Animal populations with more than, on average, 2.5 babies per individual will show chaotic behaviour, including booms and busts.

5 FALSE – The Earth's weather is governed by chaotic systems, so even with very powerful computers we will not be able to predict far into the future.

THE BLUFFER'S SUMMARY

In a chaotic system, small errors in measuring the current conditions lead to wildly different predictions about the future.

Fractals

'Clouds are not spheres, mountains are not cones, coastlines are not circles, and bark is not smooth, nor does lightning travel in a straight line.'

BENOIT MANDELBROT

The word 'fractal' was coined in the 20th century by Benoit Mandelbrot, who used it to describe shapes that are infinitely complex and self-similar. Although fractals had been studied from the 17th century onwards, it was the arrival of computers that really gave mathematicians a chance to explore the secrets of these amazing shapes, showing just how complex they could be. Exploring this new realm gave mathematicians the means to study complexities in our own world, from clouds to coastlines to cauliflowers.

Fractals are a fascinating fusion of geometry and infinity. They produce shapes with paradoxical properties, but also help us to understand complex real-world shapes.

1 A true fractal can only exist in the mathematical world and not in real life.

TRUE / FALSE

2 There is a 3-D mathematical shape that has infinite surface area but no volume.

TRUE / FALSE

3 Take a line segment. Delete the middle third. Delete the middle third of each of the two lines remaining. Keep repeating. Eventually, you will have deleted the whole line.

TRUE / FALSE

4 Measurements of the same coastlines can often differ by hundreds of miles according to different sources.

TRUE / FALSE

5 Trees and mountains in computer games are drawn in advance by artists for players to discover.

TRUE / FALSE

TEN THINGS A GENIUS KNOWS

① What a fractal is
A fractal is a shape that is self-similar at every scale. Your foot is not a fractal: zooming in on a foot does not reveal smaller feet hiding within it. But your blood vessels are fractal: zooming in on a blood vessel shows smaller blood vessels branching off, and smaller blood vessels branching off *those*. In a similar way, broccoli is a fractal because there are florets upon florets upon florets, while a carrot is not a fractal. Of course, in the real world there is a limit to how far we can zoom in, but in the mathematical world we can build fractals with infinite levels of complexity.

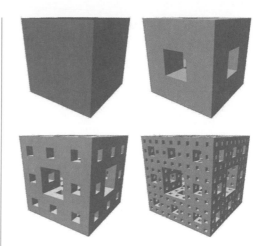

② How to build a fractal snowflake
Fractals are usually built in an iterative fashion, starting with a simple object and then modifying it over and over again. One of the earliest fractals was described by the Swedish mathematician Helge von Koch in 1904. Take

a line segment, divide it into three parts, remove the middle third, then add in to the deleted area two line segments the same size as the deleted part, forming an outward-facing triangle. Repeat the process on each of these new four smaller line segments. Repeating this process infinitely many times creates a Koch curve, and arranging three of these curves into a triangle creates a shape called the Koch snowflake.

③ What a Menger sponge is
Take a cube and divide it up into 27 equal-sized smaller cubes by dividing each side into three parts. Now remove the centre cube from each face, and remove the cube from the very centre of the large cube. There are 20 cubes left, forming a 'Level 1' Menger sponge. Now repeat the process with each of these 20 small cubes, dividing them up into 27 smaller cubes and removing all the centres. This makes a 'Level 2' Menger sponge. The true Menger sponge is created by repeating this process infinitely

many times (see image above right), creating a wondrous shape that has infinite surface area but zero volume.

④ The different ways to build a Sierpinski triangle
The Sierpinski triangle is created using a similar technique to the Menger sponge. Divide an equilateral triangle into four equal equilateral triangles, then remove the centre triangle. Repeat the process with each of the three remaining triangles, and so on to infinity. There are many other, less obvious ways to build the same fractal. One method uses the 'chaos game' and probability. Draw an equilateral triangle, label the corners A, B and C, and draw a dot at random inside the triangle. Roll a die to determine one of the corners (for example, rolling a 1 or 2 means A, a 3 or 4 means B and a 5 or 6 means C), then draw a new dot halfway between the old dot and that corner. Repeat the process with the new dot. As more and more dots are drawn inside the triangle, the shape of the Sierpinski triangle will magically appear. Another method is to draw Pascal's triangle, where each number is the sum of the two above it, and to colour in the odd numbers.

5 **The difficulties of measuring a coastline**
In 1951, the British mathematician and scientist Lewis Fry Richardson had a theory that the length of the border between two countries influenced how often they went to war. To test this theory he compiled data on the lengths of borders and coastlines. To his surprise, the data was wildly different, depending on which source he consulted, sometimes differing by hundreds of miles. He realized that the problem was the measuring stick being used: the smaller the stick, the longer the recorded length would be. This is because coastlines are not built of straight lines but are fractals, remaining highly complex on ever smaller scales. The problem of measuring coastlines is now known as the coastline paradox.

6 **How to measure the dimension of a fractal**
The French mathematician Benoit Mandelbrot invented the idea of fractal dimensions in response to Richardson's discovery of the coastline paradox. The fractal dimension quantifies how the measure of a shape changes as the stick used to measure it changes. With a normal line, making the measuring stick three times shorter means we need three times as many to fit into the length of the shape. But with the Koch curve, making the measuring stick three times shorter allows us to fit four times as many into the shape. In general, a shape has a dimension of D if a measuring stick s times shorter can fit into the shape s^D times. This gives the Koch curve a dimension of 1.26 (which is $log(4)/log(3)$), which fits our intuition that the Koch curve is a shape in between 1 and 2 dimensions. The Sierpinski triangle has a dimension 1.585 and the Menger sponge 2.73.

7 **What the Mandelbrot set is**
One of the most amazing fractals is the Mandelbrot set, which reveals an unexpectedly rich relationship between complex numbers and quadratic functions. The definition of the Mandelbrot set uses the function $f(z) = z^2 + c$, which squares a number and adds a (complex) number c. It then considers what happens if this function is applied repeatedly to the number 0. For example, if $c = 1$ then we get the sequence of numbers 0, 1, 2, 5, 26, . . ., getting bigger without bound. If $c = 0.1$ then we get 0, 0.1, 0.11, 0.1121, 0.1126, 0.1127, . . ., which gets closer and closer to a fixed number. Or, if $c = -1$,

we get 0, –1, 0, –1, 0, . . ., which oscillates between two values. The Mandelbrot set consists of those values of c where the sequence remains bounded, and when these points are drawn on the complex plane they create a picture whose intricacy must be personally investigated to be appreciated.

8 **How Julia sets relate to the Mandelbrot set**
The Mandelbrot set takes the function $f(z) = z^2 + c$, starts with the number 0 and investigates what happens as c varies. Instead, we can fix c and vary the starting number. In this way, every complex number creates its own fractal, called a Julia set (named for Gaston Julia). There is a striking relationship between the Mandelbrot and Julia sets: the Mandelbrot set is a kind of atlas, telling us exactly which Julia sets are connected (that is, all one piece) and which are disconnected, like clouds of dust.

9 **What an L-system is**
L-systems were invented in 1968 by the Hungarian biologist Aristid Lindenmayer, in order to investigate plant forms, but as they use recursion they are perfectly suited to modelling fractals. An L-system consists of a fixed set of symbols, a starting symbol, and a rule for how to go from one generation to the next. In his first example modelling algae, Lindenmayer's system had two symbols: A (adult) and B (baby). Starting with an A, the rule was that in each generation every A would become an AB (an adult stem gives birth to a baby stem) and every B would become an A (a baby would become an adult). The resulting string of characters is as follows: A, AB, ABA, ABAAB, ABAABABA, ... By interpreting these types of string geometrically, beautiful fractal drawings arise from extremely simple sets of instructions.

10 **Why computer games use fractals**
When designing a computer game, the creator can either build the entire world in advance or create it spontaneously as the player explores. As games become more complex, the second option becomes more appealing. Intricate shapes, such as mountains, clouds, trees and rocks, take a lot of computing power to produce, and would be a waste of time if the player did not go close to them to admire the detail. Instead, fractal-generating algorithms such as L-systems can be used to create such shapes in real time, adding in detail as required by the proximity of the player.

❝ In his autobiography, Mandelbrot wrote his name as "Benoit B. Mandelbrot", even though he did not have a middle name. People have speculated that the B stands for "Benoit B. Mandelbrot", thus making Mandelbrot's own name into a fractal. ❞

❝ In 2014, there was a worldwide attempt to create the world's largest fractal, a Menger sponge constructed from business cards. A single cube requires six cards, a level 1 sponge requires 120, a level 2 sponge requires 2400 and a level 3 requires 48,000 business cards. There were 20 of these level 3 Menger sponges built around the world, so it would theoretically be possible to combine them to make a level 4 Menger sponge consisting of 960,000 business cards. This project was called MegaMenger. ❞

❝ A controversial use of fractals was to analyse the paintings of Jackson Pollock. In 1999, physicist Richard Taylor claimed that Pollock's paintings had a fractal nature and that the fractal dimension of the paintings increased over his career. When a cache of 32 new paintings was discovered in an attic, Taylor was commissioned to use his fractal analysis to see if they were real or fakes, and concluded that they were all fakes. Although later work by other scientists cast serious doubt on Taylor's methods, the attic paintings are still believed to be fakes from looking at the chemicals used in the paint. ❞

1 TRUE – In real life we cannot zoom infinitely far, so there is a limit to the degree of self-similarity in an object.

2 TRUE – The Menger sponge is a fractal with infinite surface area but zero volume.

3 FALSE – Infinitely many points will still be left: the endpoints of each deleted line segment. This is called the Cantor set and is a fractal. The Menger sponge is a 3-D Cantor set.

4 TRUE – Different sized measuring sticks result in different values for the length of a coastline, because coastlines are inherently fractal.

5 FALSE – Fractal features, such as trees, mountains, clouds and rocks, are generated in real time by fractal algorithms, so as to save on computing power in the game's creation.

THE BLUFFER'S SUMMARY

A fractal is a shape that looks the same however far you zoom into it.

Unexpected trigonometry

'Mathematics compares the most diverse phenomena and discovers the secret analogies that unite them.'

JEAN-BAPTISTE JOSEPH FOURIER

What do MP3 music files, MRI scanners, quantum mechanics and a NASA Mars mission have in common? The answer is that they all rely on Fourier analysis, a technique developed in the early 19th century with no conception of how important it would become. Just as a good musician can hear the different notes in a chord, Fourier analysis can 'hear' the infinity of different notes in any signal. At the heart of the technique lie the sines and cosines we learn in school.

Nobody expected trigonometry to end up being the secret ingredient inside every piece of technology we use today – least of all the genius who figured it out.

1 The sine function describes the varying height of a point moving around the edge of a circle.
TRUE / FALSE

2 If the crest of a water wave meets the trough of another (equal-sized) water wave, then the two will cancel out and no movement will happen.
TRUE / FALSE

3 When we hear a note from an acoustic musical instrument, such as a guitar string being plucked, we are actually hearing many notes of different frequencies at once.
TRUE / FALSE

4 Sine waves cannot be used to analyse modern digital signals, because digital signals are discrete, while sine waves are continuous.
TRUE / FALSE

5 MP3 audio compression works by discarding low-frequency data in sound signals.
TRUE / FALSE

TEN THINGS A GENIUS KNOWS

❶ How trigonometry produces waves

Many of us still remember from school the various tricks for writing the formulae for sine and cosine. Tricks like SOHCAHTOA, which reminds us that in a right-angled triangle, sine is the opposite over the hypotenuse, and cosine is the adjacent over the hypotenuse. We can draw a right-angled triangle in a circle of radius 1, so that the hypotenuse goes from the centre of the circle to a point on the edge. Sine of the angle θ then describes the height of that point, and cosine θ describes its horizontal distance from the axis. Now imagine the point rotating around the circle at a constant speed. The varying height of the point over time draws out a wave, and this is called a sine wave. The speed of the rotation tells us the frequency of the wave (rotations per second), and the radius of the circle gives the amplitude of the wave.

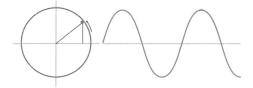

❷ How to add waves together

When two waves of any kind come together they will combine to make a new wave. The height of the new wave at each point is the sum of the heights of the two original waves. So if the crests of two equally sized waves meet, then the new wave will have a crest of double the height. But if the trough of one wave meets the crest of another wave, then these will cancel out to leave no movement at all. This addition of waves takes place whether we are dealing with water waves, sound waves, radio waves or quantum waves. In the diagram, three sine waves of varying frequencies and amplitudes are added together to create a new and very different looking wave.

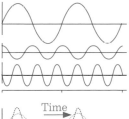
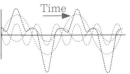

Time

❸ What Fourier analysis does

It is easy to add waves together to generate a new wave. But is it possible to go the other way? That is, to take a complicated looking wave and decompose it into a sum of sine and cosine waves? Is it possible that some shapes are so complicated that they can never be written as the sum of sine waves? Fourier analysis was designed to answer these questions. Amazingly, it turns out that any periodic function (that is, any shape that repeats itself) can be written as the sum of sine and cosine waves. Fourier worked out the formulae for how to find out which waves were needed.

❹ How Fourier analysis was used by ancient astronomers

Although the principles of Fourier analysis were laid down in 1822 by the French mathematician Jean-Baptiste Joseph Fourier, the idea went back much further. In the third century BCE, astronomers were using a version of Fourier analysis to describe the orbits of the planets in our solar system. The problem was that scientists of the day wanted a model with the Earth in the centre and the planets orbiting in circles around it, yet this did not fit what they saw. In particular, Mars would occasionally appear to move backwards in the sky – a phenomenon known as retrograde motion. To solve the problem, Apollonius of Perga proposed the idea of epicycles: that planets moved in circles that, in turn, moved in circles and so on. For a given observed orbit, the problem of deciding how many epicycles there were, along with their diameters and speeds, was equivalent to Fourier's method of decomposing a wave in terms of sine waves of different frequencies and amplitudes. (The theory was abandoned once Copernicus's heliocentric theory was accepted.)

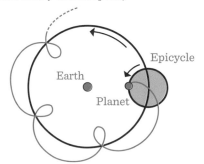

5 **What a Fourier series is**
The difference between Fourier's method and that of Apollonius was that Fourier potentially needed infinitely many sine and cosine waves to describe a wave. A Fourier series is a way of writing a periodic function $f(t)$, repeating every T seconds, in terms of an infinite sum of sines and cosines.

$$f(t) = a_0 + \sum_{m=1}^{\infty} a_m \cos\left(\frac{2\pi mt}{T}\right) + \sum_{n=1}^{\infty} b_n \sin\left(\frac{2\pi nt}{T}\right).$$

The ms and ns run through every possible frequency of wave that may potentially contribute to the sum, and the a_ms and b_ns are the amplitudes of each of these waves. Fourier had a formula to allow him to calculate these amplitudes, which involved integrating the product of $f(t)$ and the wave in question.

6 **The motivation behind Fourier series**
The reason that Fourier developed the method of writing a function in terms of sines and cosines was in order to solve the heat equation. This was a complicated differential equation describing how heat would propagate through, for example, a closed box. Before Fourier there was no way of solving this equation, because there simply did not exist a finite formula describing the heat flow. But by using Fourier's infinite sines and cosines, scientists could write down an expression for the flow of heat and do calculations with it to answer practical questions. Fourier analysis remains an indispensable method in the toolkit of mathematicians and scientists solving differential equations.

7 **The difference between a Fourier series and a Fourier transform**
A Fourier series decomposes periodic functions into sines and cosines. But not all signals are periodic: a song or a telephone conversation does not consist of a repeating series of notes. Fourier transforms extend the techniques of Fourier series to any kind of function. Instead of a summation with one term for each integer-valued frequency, Fourier transforms use an integral that sums over every possible real-valued frequency. The maths turns out to be easier if the formulae are written in terms of complex numbers, as a single complex number captures both the amplitude and frequency of a wave.

8 **How to interpret the Fourier transform**
A Fourier transform is a way to switch between thinking of a signal as a function of time to thinking of it as a function of frequency. This transformation often has a physical interpretation. In quantum mechanics, the state of a particle can be described in terms of its position or its momentum, but not both at once because of the Heisenberg uncertainty principle. The Fourier transform provides the translation between these two ways of describing a particle. In spectroscopy, the Fourier transform of a set of data describes the intensity of light at different wavelengths, which is then used to produce pictures of your body from MRI scans or to provide forensic analysis of materials.

9 **Why the fast Fourier transform is so important**
In modern technology we are usually dealing with discrete rather than continuous data. Digital signal processing samples a signal at regular time intervals to provide a sequence of data points, and in image analysis we deal with the values of individual pixels. The discrete Fourier transform (DFT) is a variation of the Fourier transform that can analyse this sort of data, but the naïve method of implementing it is too slow to be usable in practice. A technique called the fast Fourier transform (FFT), developed in 1965, is the reason that Fourier analysis is so ubiquitous, as it can compute the same answer as the DFT but in a far shorter time.

10 **How we compress data using Fourier analysis**
JPEG and MP3 compression algorithms both make use of Fourier analysis, using a special version of the DFT called the discrete cosine transform (DCT). This is the same idea as the Fourier transform, but using only cosine waves. The DCT performs a frequency analysis of the data in the picture or the music. Since the eye and the ear are bad at distinguishing between the exact strength of the high-frequency data values, these may be rounded up or even discarded without losing too much sound or picture quality. Both JPEG and MP3 can reduce file sizes by a factor of ten without being noticeable.

TALK LIKE A GENIUS

❝ As well as inventing the Fourier transform, Jean-Baptiste Joseph Fourier also has the distinction of being the first person to propose the greenhouse effect as a way of explaining the surface temperature of the Earth. He realized that the Earth was too far away from the Sun for the Sun's radiation alone to account for its temperature, and suggested that the Earth's atmosphere might be working as an insulator. ❞

❝ Fourier analysis can not only compress music and picture files but can also remove unwanted noise. This is important for projects such as missions to Mars, where noise can be a problem because the signals have such a long distance to travel before they are picked up by receivers on Earth. ❞

❝ The app Shazam allows you to record a short piece of music, such as a few bars of a song overheard in a café, and it will then identify what the music is. How it performs this magic is, in part, to find the Fourier transform of the recording and compare this to a database of musical "fingerprints". ❞

WERE YOU A GENIUS?

1 TRUE – Sine of an angle in a right-angled triangle is the length of the opposite side divided by the hypotenuse. In a circle of radius 1 this describes the height of a point on the edge of the circle.

2 TRUE – Waves will add together, so the displacement of the water particles will be the sum of the heights of the two waves.

3 TRUE – The complex tones of an instrument consist of sine waves of many different frequencies that collectively produce the quality of the note we hear.

4 FALSE – A special version of Fourier analysis, called the discrete Fourier transform, is used for digital data and in all modern signal processing.

5 FALSE – It is the high-frequency data that is discarded, as this is difficult for the human ear to detect.

THE
BLUFFER'S
SUMMARY

Fourier analysis breaks down any signal into a sum of sine and cosine waves, a technique used to analyse MRI scan data, provide data compression and solve complex equations.

Graph theory

'The optimal tour to visit the [UK's] 24,727 pubs [is] longer than the Earth's circumference, but you'll find more food and drink along the proposed route than you would during a lap around the equator.'

WILLIAM COOK

In October 2016, William Cook's research team announced the result of two years' work to find the shortest pub crawl of all the UK's pubs. The scale of the problem was a hundred times greater than that solved by anybody else, yet it is still small in comparison to the types of network problems faced by major companies every day. Graph theory is the mathematical study of networks and contains some of the hardest problems in mathematics, involving immense numbers and huge computational challenges.

Graph theory contains some devilishly difficult problems, but might the next breakthrough come from bubbles and slime mould, rather than human genius?

1 Graph theory deals with plotting and drawing data, like on a distance–time graph.

TRUE / FALSE

2 When drawing a graph, it is important to correctly measure the lengths of the edges between the nodes.

TRUE / FALSE

3 At a party with at least six guests, there will always be three who are either mutual friends or mutual strangers.

TRUE / FALSE

4 To find the shortest route that visits 100 cities, we can check all possible routes and pick the best one.

TRUE / FALSE

5 When laying down internet cables between cities, it can be more efficient to create 'phantom' towns to minimize the amount of cable used.

TRUE / FALSE

TEN THINGS A GENIUS KNOWS

1 What a graph is
When mathematicians speak of a graph, they rarely mean the graphs taught in school, such as bar graphs or distance–time graphs. A graph is a network of points, called nodes, and lines joining them, called edges. A graph is an example

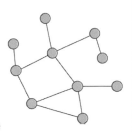

of a topological object, which means that it is not the exact placement of the nodes or the lengths of the edges that matter, but only the structure of the connections between the nodes and edges. For example, the map of the London Underground tells you how to change trains to get to your destination, but should not be used as an exact measure of distances in London.

2 How to solve the utilities puzzle
On a piece of paper, draw three houses in a line at the top, and three buildings in a line at the bottom – these are the gas, water and electricity companies. The 'utilities puzzle' asks whether it is possible to join each of the three utilities to each of the three houses using unbroken lines, in such a way that two lines never cross each other. It is an excellent puzzle to keep a friend busy for a while, because the puzzle turns out to be impossible. The graph required to solve the puzzle is called the $K_{3,3}$ graph and it cannot be drawn on a flat plane, or on a piece of paper. However, the puzzle becomes possible if the houses and utility companies are drawn on a mug with one handle!

3 Why we cannot imagine Graham's number
How many people need to attend a party in order to guarantee that there are either three guests who all know each other or three guests who are mutual strangers? The answer is six, and this question is part of an area of graph theory called Ramsey theory. But asking the same question about a party with either four mutual friends or four mutual strangers (along with some extra technical conditions – the people are modelled as corners of a hypercube and the four people must be coplanar within the cube) – results in one of the biggest numbers ever seen in a mathematical proof. Graham's number, an upper bound for the answer, is quite literally impossible to imagine: its digits contain more information than could be contained in a black hole the size of a human brain.

4 What the Chinese postman problem is
Graphs modelling real life situations often have weighted edges, meaning each edge is given a number. In the Chinese postman, or route inspection, problem, a graph traditionally represents a network of locations (houses, depots, cities) with the weighted edges giving the distances between them. The Chinese postman problem asks for a route that traverses the whole graph in the shortest possible distance, but making sure to travel over each edge at least once. This problem has a straightforward solution that can be implemented in polynomial time – that is, quickly.

5 Why the travelling salesman problem is so important
The travelling salesman problem sounds deceptively similar to the Chinese postman problem. It asks for the shortest way to traverse a weighted graph so that every node (rather than every edge) is visited. It is important in real life because of the many situations to which it applies: a company such as FedEx making deliveries in the shortest amount of time or using the least amount of petrol; programming the Hubble space telescope to view the most stars in a night; drilling circuit boards using the smallest movements of a machine; or a school bus collecting school children in the morning. But it is also important mathematically because it is part of a class of problems called NP-complete. (See page 177).These are problems whose solutions can be checked quickly, though it is unknown whether or not the solutions can be found quickly.

6 How companies solve the travelling salesman problem
One way of solving the travelling salesman problem is brute force – testing every possible solution and picking the best one. However, this is intractable even for 20 nodes. The cleverest method has solved a problem with 85,000 cities, but it took over 100 hours of CPU time to compute. The solution for the shortest UK pub crawl, whose graph had 24,727

nodes and used an existing road network, took two years to find. Since exact methods are unrealistic for companies, they instead use heuristic algorithms. They are not guaranteed to get the best solution, but give a 'good enough' solution in a short time. For example, the 'nearest neighbour' method, where the salesman picks the nearest unvisited node, usually gives an answer within about 25% of the best one. Better methods can get within 1% of the true solution while still being fast.

7 **How bubbles have mastered graph theory**
Suppose we want to install gas pipes between a collection of cities. What is the configuration of pipes that connects all the cities in the shortest distance? If the pipes need to lie along pre-existing edges (for example, under roads), then this is called the minimum spanning tree problem, and mathematicians have fast algorithms to solve it. However, if the pipes can go anywhere, the best configuration is called a Steiner tree, with additional 'phantom' nodes. Constructing Steiner trees is an NP-hard problem, which means that no fast algorithms are known. Intriguingly, bubbles often seem to find a good solution. Place nails between two sheets of Perspex in the configuration of the cities, then dunk the contraption in soapy water. The bubbles naturally form a Steiner tree between the nails, giving a good (and often the best) way to lay down the pipes.

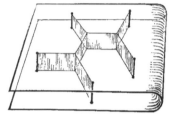

8 **How to find the shortest path across a graph**
Mathematicians are connected to each other via a publication graph, so that two people (the nodes) are joined by an edge if they have co-authored a paper. A mathematician's point of pride is their Erdős number, which is the shortest distance between themselves and the prolific Hungarian mathematician Paul Erdős in this graph. Finding the shortest distance between two nodes in a graph is important in other situations too – a satnav finding the shortest route to your destination (modelling intersections as nodes and the roads as edges), the

sequence of operations for a robot to finish a task, or protocols for sending information over the internet, for example. This is a well-studied problem, with the most widely used algorithm to solve it being Dijkstra's algorithm.

9 **How Google's algorithm uses graph theory**
The algorithm behind the ranking of Google's search results is called PageRank, a play on words between the developer Larry Page and the idea of a web page. PageRank views the web as a weighted, directed graph with web pages being the nodes. The edges are hyperlinks, going from the page where the hyperlink lives to the page that it references (hence, the edges have a direction). The idea is that, the more important a page is, the more weight the edges coming out of it have, and the importance of a page is judged by the collective weight of all the edges coming into it. Although the graph is huge, Google makes use of the structure within the graph (such as knowing that a randomly chosen pair of web pages are almost certainly not linked) to deliver reliable search results quickly.

10 **How graphs can recommend products for you**
'We see you liked this product – how about trying this other one?' This type of recommendation appears on all kinds of websites, from films on Netflix to books on Amazon, and even friends on Facebook. Recommendation algorithms can be modelled using graphs, with nodes for users and products, and edges representing engagement with the product. Different weightings are used depending on the strength of the engagement, such as a view, a like/favourite or a watch/buy. The structure of this graph is then used to make predictions of those products that a person might like, based on what they have already liked and how other similar users behaved.

TALK LIKE A GENIUS

⁶ In 2010, scientists used slime mould to model the Tokyo train network, by placing food where the train stations would be and seeing how the slime mould connected them up. They found that the slime mould network was more efficient and more robust than the real network! Similar research has now been done in other locations, including the UK, with the slime mould suggesting that the M6 be rerouted through Newcastle. ⁹

⁶ Actors have an analogy to the Erdős number called the Bacon number, which is the shortest distance to Kevin Bacon in a graph where two actors share an edge if they have been in a film together. The Erdös-Bacon number is then the sum of the two numbers. Most people's Erdös-Bacon number is infinite but a few have a number less than ten, including the scientists Carl Sagan and Stephen Hawking, and actors Natalie Portman and Colin Firth. ⁹

⁶ Mathematics helped to save the courier company UPS millions of dollars by advising their drivers not to turn left when making deliveries. The new strategy saved three million gallons of petrol in a year, and saved 32,000 tonnes of CO_2 emissions. Satellite navigation systems for UPS drivers must use a specially weighted travelling salesman problem, with left-hand turns highly weighted to discourage their use. ⁹

1 FALSE – Graph theory is about networks with nodes connected by edges.

2 FALSE – Graphs are topological objects, so the exact sizes or placement of the nodes or edges do not matter.

3 TRUE – This is a result in Ramsey theory – a theory that, at one time, created a Guinness World Record for the largest number used in a proof.

4 FALSE – Such a method is not even tractable for finding the best route between 20 cities.

5 TRUE – Graphs that connect nodes using the shortest path length often create phantom nodes, and these graphs are called Steiner trees.

THE BLUFFER'S SUMMARY

From finding the fastest way through a city to getting the best web search results, graph theory finds the answers by studying the structure of networks.

The four colour theorem

'A good mathematical proof is like a poem—this is a telephone directory.'

ANONYMOUS

The four colour theorem is one of those mathematical puzzles that is deceptively simple to state and fiendishly difficult to solve. Posed in 1852 by Francis Guthrie, it resisted the attempts of the world's best mathematicians for over one hundred years, finally being solved, in 1976, by Kenneth Appel and Wolfgang Haken. The proof is notable for being the first major mathematical result solved by using a computer, and it remained controversial for many years after publication.

A simple problem about colouring in has led to a revolution in mathematics, with computers now creating proofs that humans cannot check. Should we trust the results?

1 The four colour theorem says that no maps require as many as four colours in order for adjacent regions to have different colours.

TRUE / FALSE

2 Some maps can be coloured with only two colours.

TRUE / FALSE

3 It is possible to draw five dots on a sheet of paper and join every dot to every other dot with a line so that none of the lines cross each other.

TRUE / FALSE

4 Maps drawn on mugs can need as many as seven colours to ensure that adjacent regions have different colours.

TRUE / FALSE

5 Scheduling a set of exams can be done by colouring in a graph.

TRUE / FALSE

TEN THINGS A GENIUS KNOWS

1 The statement of the four colour problem
The four colour problem claims that it is always possible to colour the regions of a map with four colours so that no two adjacent regions share the same colour. This is a claim not just about maps of the world, or maps of countries, but about any possible drawing separating a sheet of paper into different areas. Two regions are considered adjacent if they share an edge. If they share only a corner, then they are allowed to be the same colour.

2 How to restate the four colour problem in terms of graphs
Mathematicians always strip problems down to the bare minimum of information needed to solve them. In deciding how to colour a map, the sizes of each region are irrelevant – these can therefore be shrunk down to dots (or nodes). We connect two nodes by an edge if the two corresponding regions in the original map were adjacent. The four colour problem then becomes: can we colour the nodes of the resulting graph using four colours so that whenever two nodes are connected by an edge they are different colours?

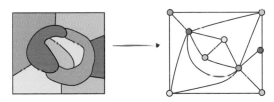

3 Why four colours are interesting
Why do we have the four colour problem and not the three colour problem or the five colour problem? It turns out to be easy to find an example of a map that cannot be coloured with only three colours. The drawing above is one such example, because in the middle are four regions that are all adjacent to each other. Going the other way, mathematicians quickly found a proof that all maps could be coloured using five colours. This was done in 1890 by Percy John Heawood, who based his five-colour proof on a failed four-colour proof by Alfred Kempe. But the four colour problem remained a mystery, with mathematicians neither being able to find a proof nor a counter-example.

4 What makes the four colour theorem difficult
The four colour theorem feels like it should be simple, yet an hour spent colouring in some maps will quickly show the difficulty. It can happen that part of a map is successfully four-coloured, but then a new region makes further colouring impossible. For example, if the outside region of the drawing above had to be coloured, it could not be done consistently with the existing colouring, because four different colours are already used on the boundary. In this case, it can be easily fixed: change the black region on the right-hand side to the lightest grey and you can colour the outside region black. But, in general, there may need to be a whole sequence of changes made, in order to accommodate the new region.

5 Why Kempe's proof failed
Kempe's 1879 proof of the four colour theorem was incorrect, but the basic idea formed the foundation of the later successful proof by Appel and Haken. Kempe worked on a proof by contradiction: he assumed that there were maps complex enough to need a five-colouring, picked the smallest example of one and called it F. Using Euler's formula (see page 96), he first showed that within F there must be at least one region with five or fewer neighbours. Removing this region created a smaller map, for which a four-colouring was possible. Kempe then tried to add the region back in and systematically recolour the graph to create a global four-colouring, contradicting the fact that F needed five colours. But the recolouring method failed in the case where the region had five neighbours.

6 How Kempe chains work
Kempe chains were Kempe's method of recolouring maps to accommodate the region that had been removed. If the region in question had exactly four neighbours, then it is possible those neighbours were all given different colours in the four-colouring of the smaller map. Say the colours were blue, yellow, red and green going round in a circle. Kempe's idea was to follow chains of blue/red or yellow/green regions around the map and, in a clever way, reverse the colouring on one of these chains while still keeping a four-colouring. This meant that the deleted region was only surrounded by three colours, leaving a fourth colour free for itself.

7 Why computers were needed to prove the four colour theorem

Kempe chains fail to work when a region has five neighbours, so a new technique was needed. Instead of removing a single region, mathematicians tried to remove a whole collection of regions. They would then recolour the map with four colours, add the collection of regions back in and extend the colouring. If such an extension of the colouring was possible, the collection of regions was called a reducible configuration. The idea was to find a finite collection of such configurations and to show that every map had to contain one of them, because then the procedure would result in a four-colouring for every map. Computers were used in the part of the proof that extended the four-colouring, in order to check all possible re-colourings and find one that extended to the deleted configuration.

8 The controversy over the final proof

In the final proof by Appel and Haken, there were over 400 pages of handwritten proof and 1936 reducible configurations that computers had checked could work. Mathematicians could check the handwritten part, but the community was suspicious of the computer part of the proof. Although different programs double-checked the result, the computations were so long, that a human could not possibly verify them. In 1996, mathematicians Robertson, Saunders, Seymour and Thomas reduced the number of configurations to be checked down to 633, but these could still not be checked by hand.

9 The role of snarks

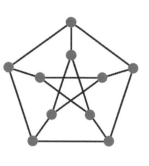

A snark is a graph in which every node has three neighbours, but that cannot be coloured with three colours so that each node has three different coloured edges coming out of it. In 1880, the Scottish mathematician Peter Guthrie Tait showed that the four colour problem would be solved if he could show that every snark drawn on a sheet of paper would have at least two edges crossing each other. William Tutte conjectured that all snarkiness was due to the Peterson graph (above) – that is, that any snark contained the Peterson graph within it. Since it was known that every flat drawing of this graph had two lines crossing, Tutte's conjecture would prove the four colour theorem. In 2001, Robertson, Saunders, Seymour and Thomas proved this conjecture, but still relied on a computer for part of the proof.

10 The uses of graph colouring

Graph-colouring techniques have been applied to a number of practical problems. The most common application is in scheduling, where we wish to organize a number of events (for example, exams, rotas, manufacturing jobs), but there are restrictions on which ones can happen simultaneously (a student may need to attend two different exams, or two jobs may both require the same resources). The events are modelled as nodes in a graph and an edge is drawn between them if they are in conflict. The minimum number of colours needed so that an edge doesn't connect two nodes of the same colour is a measure of the minimum time needed for all the events to take place.

❝ People have questioned the validity of the proof of the four colour theorem because it cannot be verified by a human. While the code itself can be checked, it may be that a subtle problem in the computer architecture, such as how it deals with rounding numbers, may cause errors to happen when the code is executed. But is this really less reliable than a human proof? Kempe and Tait's proofs were both published in respectable journals and stood for 11 years before mistakes were found. ❞

❝ Although snarks were invented in 1880 they were not named until 1976, when Martin Gardner called them snarks after the mysterious creature in the Lewis Carroll poem *The Hunting of the Snark.* ❞

❝ To bamboozle a friend who knows about the four colour theorem, draw a map on a mug and ask them to four-colour it. In fact, the four colour theorem only applies to graphs drawn on flat paper (or a sphere) and there are maps on (one-handled) mugs that require seven colours. ❞

1 FALSE – The four colour theorem says that all maps can be coloured with four colours so that adjacent regions have different colours.

2 TRUE – Some maps need just two colours, and others require only three, so that adjacent regions have different colours. But there are some maps that require four.

3 FALSE – If such a thing were possible, there would be maps that need five colours, owing to the relationship between maps and graphs.

4 TRUE – Topologically different objects require different numbers of colours for map colouring, and for a mug (torus) the number is seven.

5 TRUE – Scheduling problems can be modelled using a graph, with the minimum number of colours giving the minimum amount of time for the events to take place.

THE BLUFFER'S SUMMARY

The regions of any map can always be coloured using four colours, so that adjacent regions have different colours.

P versus NP and the problem of fast algorithms

'If P = NP then everyone who could appreciate a symphony would be Mozart, and anyone who could recognize a good investment strategy would be Warren Buffet.'

SCOTT AARONSON

P versus NP is one of the Millennium Prize Problems worth a million dollars. At its heart, it asks whether problems with solutions that are easy to check are also easy to solve. The majority of mathematicians believe the answer is no, but a positive answer might have serious consequences for many areas of our lives – from encryption on our credit cards to the delivery of our post. Despite over a hundred purported proofs, no genius has yet solved the problem.

Checking the solution to a puzzle is easy – it's finding the solution in the first place that is hard. But might the two ideas be related? One million dollars is yours if you're smart enough to resolve the question.

ARE YOU A GENIUS

1 Deciding whether a number is a multiple of 5 gets harder as numbers get bigger.
TRUE / FALSE

2 The problem of putting a list of numbers in order, from smallest to largest, can be solved in polynomial time (quickly).
TRUE / FALSE

3 I am thinking of a number between 1 and 100. Asking only yes/no questions, you need at least ten questions to guess my number.
TRUE / FALSE

4 NP stands for 'non-polynomial', and is the class of problems not solvable quickly.
TRUE / FALSE

5 If a proof of P = NP were found, we would immediately have fast algorithms to break all encryption.
TRUE / FALSE

TEN THINGS A GENIUS KNOWS

❶ Different types of complexity of an algorithm

The complexity of a computer algorithm is the time it takes to run, based on the input to the program. As an example, consider the problem of deciding if a number is odd or even. The method is to look at the final digit of the number in binary: if it is 0 then the number is even, and if it is 1 then the number is odd. The complexity of this procedure is constant: it does not depend on the input. Another example is the problem of finding the largest value in a list of numbers. The procedure is to go through the list, one number at a time, keeping track of the largest number found so far until the last number is reached. The running time for this algorithm is linearly dependent on the number of values in the list – if we double the length of the list, the program takes twice as long to finish.

❷ What the 'big O' notation is

Mathematicians use the notation of 'big O' to mean the time complexity of an algorithm. Finding the largest number in a list of values has $O(n)$ time complexity, which means that the time taken for the algorithm to finish is of the same order as the number of inputs n. If we were not only to find the largest number in the list, but to put the list in order from smallest to largest, we could use a method called 'bubble sort', which keeps passing through the list, finding the next highest number at each pass. This algorithm has complexity $O(n^2)$ – the time taken to finish increases as the square of the number of inputs, so if there are twice as many numbers in the list, the method takes four times as long.

❸ What it means to be a polynomial time algorithm

Saying that an algorithm finishes in polynomial time means that the time complexity of the algorithm is not larger than a power of the number of inputs. Bubble sort is a polynomial time algorithm because it grows like n^2, which is a power of the number of inputs n. A less obvious example is the binary search algorithm, which guesses the number you are thinking of by halving the search space each time. If you are thinking of a number between 1 and 100 it can find the answer within seven steps. This algorithm has complexity $O(log\ n)$, where $log\ n$ is the logarithm of n. The logarithm function grows more slowly than any power of n, so is at most polynomial time. On the other hand, suppose we were trying to guess a particular ordering of the numbers 1 to 100. There are 100 factorial ($100 \times 99 \times 98 \times ...$) combinations, so guessing each one in turn is an $O(n!)$ algorithm, which is exponential instead of polynomial time.

❹ What P and NP are

The set P consists of problems whose solutions can be *found* in polynomial time. For example, the bubble sort and binary search algorithms are in the class P. The set NP consists of problems whose solutions can be *checked* in polynomial time. Every algorithm that is in P is also in NP, but not every algorithm in NP is necessarily in P. The puzzle in which we had to guess an ordering of numbers is naively an exponential time algorithm of $O(n!)$, but checking the solution is only $O(n)$ because we just compare each number in our answer against the true answer to see if it is right. The problem worth a million dollars, called P vs NP, is to decide whether or not the sets P and NP are the same. Does having a fast check imply having a fast way to find a solution?

❺ What NP stands for

NP stands for 'non-deterministic polynomial time'. An equivalent definition of the class NP is that it is the set of problems whose solutions can be found in polynomial time, using a non-deterministic (Turing) machine. A non-deterministic computer is one that, for each possible input, may have multiple different outputs. This leads to a branching tree structure where we imagine that all branches are being investigated at once. If a problem is in NP, then there are polynomially-many steps to getting the right answer, which means that the correct route through the tree can be checked in polynomial time.

❻ Why NP-complete programs are important

There is a class of problems called NP-hard. A problem in this class is at least as hard as any NP problem. This means that any problem in NP can be reduced to one that is in NP-hard. If a problem is both in NP and in NP-hard, we say it is NP-complete. So an NP-complete problem is, in some sense, one of the hardest problems out of all the NP problems. The set of NP-complete problems is important because it provides a route to deciding whether P = NP: if even a single problem in NP-complete can be solved in polynomial time, then so can all NP problems.

7 **Examples of NP-complete problems**
Probably the most famous NP-complete problem is the travelling salesman problem (see page 169) in which a salesman must visit every node of a graph at least once. A particular version of the problem asks whether there is an answer taking less than a certain amount of time or distance. The answer to this is easily checked, but we do not yet know any polynomial-time method for getting the answer in the first place. Another important example is the knapsack problem, where a traveller must pack a knapsack with items to get the largest possible value out of the bag. (The knapsack problem could also apply to a shop deciding which products to put on its shelves to maximize profits.)

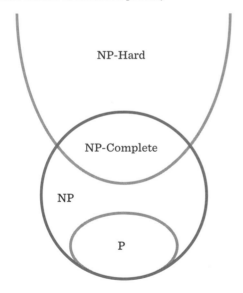

8 **Whether integer factorization is NP-complete**
Finding the factorization of a number such as 3,322,127 would take a large amount of time, but if I claim that the answer is 1171 × 2837 then you can check whether I am correct very quickly. So integer factorization is a member of NP. The best algorithm known to solve it has $O(2^{3\sqrt{n}})$, which is called sub-exponential and is larger than polynomial

time. But even if a polynomial-time method were found for solving the problem (which in itself would be catastrophic for most encryption methods), this would not solve the P vs NP problem because it is not known whether integer factorization is in NP-complete. That is, knowing that the problem is in P would not show that every NP problem is in P.

9 **What the consequences of P = NP would be**
It is common to see claims that a proof of P = NP would change the world as we know it. It would mean that there would be 'fast' methods for breaking encryption, for solving difficult logistic problems and even for inventing programs to prove mathematical theorems. But 'fast' is a matter of perspective. A polynomial time method may have very large exponents and so will only be faster than other algorithms when the inputs are incredibly large. For all practical purposes, an exponential algorithm with small coefficients would perform better. Added to this caveat, a proof that P is equal to NP may not be constructive, in the sense that it may provide no practical insights into how to find polynomial-time algorithms in the first place.

10 **Whether mathematicians believe that P = NP**
In a 2012 poll of 152 mathematicians, 83% believed that P ≠ NP, 9% believed that P = NP, and 3% believed that the answer could not be resolved from the current axioms of mathematics. It is possible that there are no polynomial-time algorithms for NP problems, but that this result cannot be proved from the axioms of mathematics, or that polynomial-time algorithms can be found, but they cannot be proven to work.

❦ The question of whether P equals NP is in some sense self-referential. If there were a proof of P = NP that could be checked in polynomial time, then the P = NP problem is itself in NP. So the difficulty of finding a proof could depend on the proof itself. ❧

❦ In the film *Travelling Salesman*, the US government hires four brilliant mathematicians to solve P versus NP, in order to break all cryptographic systems. The focus of the film is the morality of giving this proof to the government while keeping it secret from the rest of the world. ❧

❦ Quantum computers sound similar to non-deterministic computers, in that they can investigate multiple different states simultaneously. But they are not the same thing, and the invention of a quantum computer will not necessarily mean we are able to quickly solve NP problems. However, some quantum algorithms have already been designed for specific problems, such as integer factorization. ❧

1 FALSE – We can easily decide if a number is a multiple of 5 by looking at the last digit.

2 TRUE – This can be solved using an algorithm called 'bubble sort', which scales as the square of the number of values in the list.

3 FALSE – You need only seven questions to guess the answer, using a method of halving the search space each time.

4 FALSE – NP stands for 'non-deterministic polynomial time', and is a class of problems whose solutions can be checked quickly.

5 FALSE – Polynomial-time algorithms may not necessarily be faster for practical purposes than exponential ones.

THE BLUFFER'S SUMMARY

It is not known whether a problem whose solution is quickly verified can be quickly solved, but the consequences will be important for everything from commerce and logistics to data encryption.

Cellular automata

'If people do not believe that mathematics is simple, it is only because they do not realize how complicated life is.'

JOHN VON NEUMANN

Can a machine make replicas of itself? This was the question that led to John von Neumann and Stanislaw Ulam inventing the concept of cellular automata, which went on to capture the public imagination with John Conway's Game of Life in the 1970s. This is the genius idea that can simultaneously explain patterns on seashells, voting trends in communities, a plant's regulation of carbon dioxide and the spread of epidemics. Cellular automata may even be able to answer our most fundamental questions about the universe and life itself.

Could the complex dynamics of the world around us arise from just a small collection of simple rules?

1 In a line of people where each person sits or stands depending on what their two neighbours are doing, all possible rules lead to simple and repetitive behaviour.

TRUE / FALSE

2 It is impossible to create a machine that will replicate both itself and the instructions for replication.

TRUE / FALSE

3 If people are influenced in their voting by what their neighbours are doing, this always leads to communities segregated by political party.

TRUE / FALSE

4 The stomata on plants (which regulate carbon dioxide intake) decide whether to open or close, depending on what neighbouring stomata are doing.

TRUE / FALSE

5 In our modern laws of physics, each particle is influenced only by the other particles that are closest to it.

TRUE / FALSE

TEN THINGS A GENIUS KNOWS

1 What a cellular automaton is

In a cellular automaton, there is a collection of cells, with each one in a particular state (of which there are finitely many). The cells are usually considered to be square and arranged in a grid but, in theory, could be any shape and arranged in any configuration. For example, the cells could represent bacteria that are either alive or dead, or they could be pixels on a screen that are red, green or blue, or they could be people supporting different political parties. For each system there is an agreed 'neighbourhood' of a cell. The state of a cell in the next time step is determined by the states of the cells in its neighbourhood, according to a fixed rule that applies to all cells simultaneously.

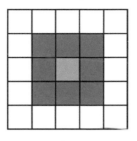

2 How they work in one dimension

The simplest type of cellular automaton is a one-dimensional version where the cells lie along a single line and are each in one of two states, labelled 0 or 1. How many different rules are there for how these cells could evolve? The state of each cell in the next time step will depend on its current state and the states of its two neighbours: the cell on its left and the cell on its right. For every possible combination of 0s and 1s that might make up these three cells, we must specify either a 0 or a 1 for what the middle cell will be in the next time step. These can be written out in a grid like the one below. Since there are two options for each of the eight states, this leads to $2^8 = 256$ different rules for evolution. Stephen Wolfram developed a notation for these rules by taking the sequence of eight 0s and 1s, thinking of it as a binary number (see page 12) and then converting it to a decimal. So in the example shown, the rule is 01011010, which is Rule 90.

111	110	101	100	011	010	001	000
0	1	0	1	1	0	1	0

3 The different types of cellular automata behavior

Even 1-D cellular automata can display an amazing variety of behaviour. This is best seen by drawing out the evolution of the system on a 2-D grid, using the second dimension as time. Some rules, such as Rule 44, eventually settle down to a steady state where the cells no longer change. Other rules, such as Rule 152, develop into an oscillating system where the cells start repeating a sequence of behaviour. More interesting than either of these two cases are the chaotic rules. These result in seemingly unpredictable behaviour, where changing the value of even one cell at the beginning of time results in a completely different evolution of the system. Rule 30 is a classic example of this behaviour. A fourth class of rules sits on the 'edge of chaos', like Rule 110, where structures and patterns seem to form but in highly complex ways.

| Rule 44 | Rule 152 | Rule 30 | Rule 110 |

4 What the Game of Life is

Unsurprisingly, 2-D cellular automata are vastly more complex than 1-D automata. The best-known example is The Game of Life, which uses a square grid extending out to infinity in each direction. Each cell is either alive (black) or dead (white) and its neighbours are the eight cells immediately touching it on the sides and diagonals. The Game of Life was invented in 1970 by John Conway, who wanted to find a simple set of rules that could create complex unpredictable phenomena, with the potential to create self-replicating machines as von Neumann had envisioned.

5 The rules of the Game of Life

The evolution of the Game of Life follows five rules. First, if a cell is alive and has exactly two or three live neighbours, then it stays alive. Second, if a live cell has fewer than two live neighbours, then it will die from loneliness/underpopulation. Third, if a live cell has more than three live neighbours, then it will die from overpopulation. Fourth, if a cell is dead but has three live neighbours, then it will become alive in the next round, from reproduction. Finally, if a cell is dead and does not have three live neighbours, then it will stay dead.

6 How the Game of Life can emulate a computer

Given the simple rules that define the Game of Life, people have discovered an amazing number of complex structures within the system. Just like the 1-D automata, there are shapes that do not change once they are created, and other patterns that oscillate between a fixed number of states. In addition, there are 'spaceships' that move across the grid in straight lines, 'guns' that fire off spaceships, and 'breeders' that travel across the grid creating guns as they go. Combinations of these structures, including spaceships, can be used to construct logic gates and thereby turn the Game of Life into a computer. It actually becomes a universal Turing machine (see page 61), meaning that it is as powerful as any other computer in existence.

7 What a universal constructor is

The original motivation for the construction of cellular automata was to create a machine that could make copies of itself. More than that, von Neumann wanted to create a 'universal constructor' that could build anything at all, given the correct instructions. In the 1940s, he proved that a 29-state cellular automata starting with 200,000 cells would be capable of such a thing, but it was unknown whether the much simpler Game of Life could achieve this. It took until 2013 before the first true replicator was built – one that could copy both itself and the instructions to repeat the process. The construction is due to Game of Life enthusiast Dave Greene, and is massive. If each cell were 1mm wide then Greene's replicator would start out being 15km across.

8 How nature uses cellular automata ideas

Scientists have found that nature often acts like a cellular automaton, with cells performing certain actions depending on what their immediate neighbours are doing. Pigment-producing cells on seashells act in this way, with the result that complex and seemingly random patterns can be modelled by simple cellular automata rules. The shell *Conus textile,* for example, bears an uncanny resemblance to the patterns seen in Rule 30. Plants use a version of cellular automata to make sure they get the most carbon dioxide while losing the least amount of water. Openings called stomata seem to open and close depending on what their neighbours are doing, and patches of open and closed stomata appear to move around a leaf in a way reminiscent of the spaceships of the Game of Life.

9 How cellular automata model real life

Many aspects of our world can be modelled using 2-D cellular automata. One example involves our voting habits. Suppose that a person will vote for the party that the majority of their neighbours vote for. Researchers have shown that, even if a community starts out being completely mixed in its voting preferences, the situation will quickly evolve to have large areas segregated by political party. Cellular automata have also been indispensable in modelling how epidemics and forest fires spread. Both infections and fires will spread if there is a high enough population density of people or trees, but even simple models can allow scientists to vaccinate the right people or burn off the right areas in order to prevent an epidemic or natural disaster.

10 Whether cellular automata may be able to model the universe

The Game of Life is defined by a small number of simple rules, but can demonstrate astonishingly complex behaviour, including lifelike moving structures and self-replicating areas. Is it possible that the complex behaviour we see in the world around us could be explained by a higher-dimensional cellular automaton with a collection of simple rules? A number of people have investigated this possibility, but most scientists remain sceptical. One stumbling block is Bell's theorem, which says that no theory involving local deterministic rules can reproduce the predictions of quantum mechanics. A cellular automaton with non-local rules might provide a loophole, but more work is needed before it becomes a serious contender for a 'theory of everything'.

TALK LIKE A GENIUS

❝ John von Neumann's vision of a universal constructor predated the discovery of the mechanism of DNA. Just like his self-replicating machine, a strand of DNA contains the instructions for replicating an exact copy of itself, leaving behind the original version. ❞

❝ Rule 90 is created by a simple rule: if the cells on either side have the same value as each other, the middle cell becomes a 0; if the two cells have different values, then the middle cell becomes a 1. When seeded with a random initial configuration, the evolution seems to be chaotic, yet when seeded with a row of 0s containing a single 1, the evolution produces an amazing fractal: the Sierpinski triangle. ❞

❝ In 2017, a new train station was opened named 'Cambridge North'. The outside of the building is decorated with beautiful panels displaying a cellular automata pattern. The designers apparently wanted to showcase images from Conway's Game of Life, as the concept was developed while Conway was at the University of Cambridge. However, the designers made a mistake in their research and instead created their panels based on Wolfram's Rule 30. The only problem: Wolfram went to rival university Oxford. ❞

1 FALSE – Some rules can lead to highly complex and even chaotic behaviour.

2 FALSE – Von Neumann created such a virtual machine in the 1940s, and a replicator in the Game of Life was discovered by Dave Greene in 2013.

3 TRUE – Simple cellular automata models predict this kind of segregation, even when communities are completely mixed in their voting preferences to start with.

4 TRUE – Stomata open and close in response to the behaviour of their neighbours.

5 FALSE – Modern physics has shown that quantum mechanics is incompatible with theories using only local deterministic rules.

THE BLUFFER'S SUMMARY

Cellular automata are used to model situations, like voting behaviour and forest fires, where the state of each individual changes according to the states of their neighbours.

Game theory

'Game theory was developed by a genius and assumes that other people are geniuses.'

TIM HARFORD

John von Neumann was a genius from a young age. His parents would bring him out at dinner parties to multiply eight-digit numbers in his head, converse in ancient Greek or recite memorized pages from the phone book. In 1928, he created a new area of mathematics called 'game theory'. But von Neumann was not the only genius to work in this field. John Nash earned his PhD with a brilliant 28-page dissertation on game theory and his work later went on to win a Nobel Prize in Economics. Game theory has had an incredible impact, generating ten other Nobel prizes, but is often criticized for assuming that the protagonists in a game act perfectly rationally when, in reality, humans are just as likely to do the opposite of what mathematics predicts.

Decision-making is fraught with difficulty – everybody tries to second-guess everyone else. Can mathematics help?

1 A payoff matrix is a grid of numbers showing the payoffs for every combination of strategies in a game.

TRUE / FALSE

2 A zero-sum game is where one player's gain is another player's loss.

TRUE / FALSE

3 Sometimes, even if you lose, the best option is to play the same strategy again next time.

TRUE / FALSE

4 In some games, the best way to play can be to choose a strategy completely at random.

TRUE / FALSE

5 Game theory is very good at modelling human behaviour.

TRUE / FALSE

1 The basic ideas of game theory

Game theory is an attempt to model the behaviour of participants competing or cooperating for resources. It studies the payoffs for each player under different strategies, based on the strategies used by the other players. This is best illustrated by an example. In the game of Chicken two players are driving cars towards each other on a straight road and the idea is to see who swerves first. If both players swerve at the same time the game is a tie; if one player swerves but the other does not, then that player is called a chicken and is ridiculed while the other wins; but if neither player swerves then both will be seriously injured in the crash. The choice of what strategy to take – stay straight or swerve – depends on what you think your opponent will do.

2 How to analyse a game of chicken

The different outcomes in a game of Chicken can be modelled in a payoff matrix. This is a table of numbers that has the strategies for Player 1 along the rows, the strategies for Player 2 along the columns, and values in the grid showing the respective payoffs for each player for each combination of strategies. The grid below represents such a possible payoff matrix. It is worth noting that only the relative values are important. That is, the outcome where both players stay straight is worse than being a chicken, being a chicken is worse than a draw, and winning against a chicken is better than a draw.

	Swerve	Straight
Swerve	0 , 0	-2 , +2
Straight	+2 , -2	-10 , -10

3 What a zero-sum non-cooperative symmetric game is

The game of Chicken is symmetric: if the players were to switch sides, the payoff matrix would remain the same. If, instead, one of the cars was specially reinforced so that it would not sustain much damage in a crash, then the result of both cars staying straight might be (-10,-3), and the game

would no longer be symmetric. Chicken is also a non-cooperative game, because the players cannot make binding commitments as to their strategies. The players might agree to swerve at a mutually agreed point, but nothing commits them to this action except their word. If they signed a legally binding contract agreeing to a course of action, it would become a cooperative game. Finally, chicken is not a zero-sum game, because the sum of the payoffs in each outcome is not zero. In a zero-sum game, one player's win is another player's loss, but in Chicken both can lose.

4 What a perfect information game is

A game of perfect information is one where every player knows all the moves made by previous players. A game like Chicken, where both players play simultaneously, cannot be a game of perfect information, but in a game like chess, where play is sequential, it is possible for all players to share the same information when making a move. Card games such as poker are sequential, but they are also not perfect information, because when a player picks up a card they do not share the information of what is on the card with the other players. Sequential games are more often represented by a tree than by a matrix, as it becomes important to know the order in which decisions are made.

5 What a dominant strategy is

If a player is better off playing a particular strategy, no matter what the other player does, then this strategy is said to be dominant. The payoff matrix below represents a game in which each person has to think of a whole number between 1 and 10.

	Even	Odd
Even	1 , 3	3 , 2
Odd	-1 , 2	2 , 2

If Player 2 picks an even number, Player 1 is better off picking an even number because they win 1 instead of -1. If Player 2 picks an odd number, then Player 1 is still better off picking an even number, because they win 3 instead of 2. So playing 'even' is a dominant strategy for Player 1. Similarly, playing 'even' is also a dominant strategy for Player 2, although it is only weakly dominant because if Player 1 played 'odd' then Player 2 would have nothing to choose between the two options.

6 What a Nash equilibrium is
A Nash equilibrium is a collection of strategies (one for each player) from which no individual player can benefit from changing their own strategy. For example, (even, even) is a Nash equilibrium in the previous game, because if both players know that the other is playing 'even', then neither has an incentive to switch to playing 'odd'. Dominant strategies, if they exist, will always be a part of a Nash equilibrium, but not every equilibrium need involve a dominant strategy. There may also be more than one Nash equilibrium in a game. In Chicken there are two equilibria: the (swerve, straight) or the (straight, swerve) strategies. If, on the first round, Player 1 swerved while Player 2 stayed straight then in the second round both players would do the same again if they each assumed that the other player would do the same again.

7 Why a mixed strategy is sometimes best
It is possible to have a Nash equilibrium that is a mixture of different strategies. The classic game of Rock, Paper, Scissors does not have any 'pure' strategies, because every choice can always be defeated by another choice. The best thing to do is to play randomly, picking each strategy with probability 1/3. In the game of Chicken, there is also a mixed Nash equilibrium in addition to the two pure strategies already seen. In the mixed strategy, the players choose randomly between staying and swerving, with the exact probabilities depending on the expected payoffs in each circumstance. John Nash proved that, in any game with a finite number of players, each having a finite number of choices, and allowing for mixed strategies, there will be at least one Nash equilibrium.

8 How to model cooperative behaviour
There are many examples in which animals or humans working together can achieve better outcomes than working alone, but only if every individual plays their part. A simple version of this scenario is modelled in game theory by the 'stag hunt'. An individual can choose to hunt either a stag or a hare, but they can only get the stag if the other player

chooses to hunt it with them. There are two pure Nash equilibria: (hare, hare), which minimizes risk, and (stag, stag), which maximizes payoff. The likely outcome will depend on the level of trust that the two players have for each other. The two Nash equilibria show that lost trust is hard to regain, but also that gained trust reinforces itself.

	Stag	Hare
Stag	3 3	0 1
Hare	1 0	1 1

9 How game theory is used in biology
Evolutionary Game Theory is an important area of study in biology for describing why animals behave or look the way they do. The idea is that strategies that perform better result in higher numbers of offspring and so the genes responsible for that behaviour will be amplified in the population. In the animal world, the game of Chicken is called Hawk-Dove, and is used to model two animals competing for a resource. 'Hawk'-like animals are aggressive and will fight until they are dead or injured, while 'dove'-like animals will try to share a resource; they can display threatening behaviour but will run away if attacked. The percentages of the two behaviours in the population will reach an 'evolutionarily stable strategy' similar to the mixed Nash equilibrium predicted by game theory, since an excess of hawks or doves makes it wiser for an animal to adopt the reverse strategy, re-stabilizing the population.

10 The ways in which humans aren't rational
In many games, people do not behave as game theory predicts they should. A classic example is the Ultimatum game. Here Player 1 is given, say $100, and they are invited to share some of this with Player 2. If Player 2 accepts the deal, they each get the amount of money they agreed on, but if Player 2 rejects the deal, they both get nothing. Mathematically, the strategy is obvious: Player 1 should offer the least (positive) amount of money possible, reasoning that Player 2 would rather accept something than nothing. But real-life experiments show that this strategy is rarely used. Offers of $50 or more are common, and offers under $30 are often rejected, even when the game is once-only and between strangers.

TALK LIKE A GENIUS

❝ The life story of John Nash was made into the film *A Beautiful Mind*, featuring Russell Crowe. It focuses on his battle with paranoid schizophrenia. John Nash and his wife Alicia died in 2015 after the taxi they were in crashed into a guardrail. They were on their way home from a trip to Norway where Nash had received the Abel Prize, one of the highest honours in mathematics. ❞

❝ The side-blotched lizard exists in three forms: orange, blue and yellow. Just as in a game of Rock, Paper, Scissors, the sneaky yellows dominate the oranges, the aggressive oranges dominate the blues, and the monogamous blues dominate the yellows. When John Maynard Smith, the biologist who invented the concept of the evolutionarily stable strategy, heard about the lizards he exclaimed "They have read my book!" ❞

❝ In the Ultimatum game, the choices we make can be influenced by various hormones and drugs. Increased levels of oxytocin (which controls empathy) make people more generous in their offers as Player 1, while people with low serotonin (which affects well-being and happiness) are more likely to reject unfair offers as Player 2. People intoxicated with alcohol were also more likely to reject unfair offers. ❞

WERE YOU A GENIUS?

❚ TRUE – A payoff matrix contains all possible outcomes of a game and the respective payoffs to each player for each one.

❷ TRUE – In a zero-sum game, the sum of all the players' payoffs is zero, so if one player wins then another player must lose.

❸ TRUE – For example, in a game like Chicken, (swerve, straight) is a Nash equilibrium, meaning that both players would choose the same strategies if played a second time, assuming the other player is doing the same.

❹ TRUE – This can happen in games such as Rock, Paper, Scissors, where every strategy can be defeated by another strategy.

❺ FALSE – Humans often behave in a non-rational manner, going against the predictions of game theory.

THE BLUFFER'S SUMMARY

In situations in which individuals compete for resources, whether that be meals, mates or money, game theory can help to find the best strategies for winning.

The prisoner's dilemma

'The best way to find out if you can trust somebody is to trust them.'

ERNEST HEMINGWAY

The prisoner's dilemma is one of the most important and confounding examples in game theory. It is a paradox of how individuals can make perfectly logical choices about what is best for themselves, yet achieve a worse outcome than if they had worked together as a group. Its analysis has been used to understand many real-world situations, from petty theft to advertising wars, and from doping in sport to the difficulties of global climate agreements. It has also been the basis of many a game show and film plot.

The prisoner's dilemma explains fundamental human traits, such as our propensity for both revenge and forgiveness, and is a perpetual source of fascination to economists, sociologists and biologists alike.

1 Two shops in a street will always be tempted to undercut one another's prices, even though doing so reduces profits for both.

TRUE / FALSE

2 Choosing strategies that maximize our own winnings is always the best way to come out ahead in a game.

TRUE / FALSE

3 The best strategy for a game can change, depending on whether you are playing it once or playing it multiple times.

TRUE / FALSE

4 In games of mutual cooperation, taking revenge on traitors is not a helpful strategy.

TRUE / FALSE

5 Being kind, forgiving and non-envious are traits that help groups of individuals to succeed.

TRUE / FALSE

TEN THINGS A GENIUS KNOWS

① **What the prisoner's dilemma is**
The prisoner's dilemma is most commonly stated in terms of a scenario involving two gangsters brought in by the police. They are taken to separate cells and prevented from communicating in any way. The police have enough evidence to convict each of them for a year on petty crimes, but they need more evidence to convict them of a big heist. They offer the prisoners a deal: if one prisoner defects and provides evidence while the other prisoner does not, then the police will let the defector go free while giving the other prisoner five years for obstructing justice. However, if both prisoners confess, then both will get three years in jail. Should the prisoners cooperate with their partner and stay silent, or defect and confess to the police?

② **How to analyse the prisoner's dilemma**
The prisoner's dilemma can be represented in a payoff matrix where the number of years in jail are the (negative) payoffs:

	Cooperate	Defect
Cooperate	-1 , -1	-5 , 0
Defect	0 , -5	-3 , -3

Prisoners looking at the options would think to themselves: 'If my partner cooperates with me and stays silent, then I have more to gain by defecting, getting out of jail free instead of serving a year. If my partner defects and confesses to the police, then I am better off defecting too, since I will get only three years in jail instead of five. Therefore, I am always better off defecting, no matter what happens.' The paradox of the prisoner's dilemma is that if the prisoners follow this logical reasoning, doing what is best for themselves, they both end up in a worse position than if they had cooperated.

③ **Why a Nash equilibrium is not always optimal**
The optimal outcome of the prisoner's dilemma is mutual cooperation. This strategy minimizes the total number of years in jail for both players, and no other strategy can improve the outcome for one player without making it worse for the other. But mutual cooperation is not a Nash equilibrium, because if one player is cooperating, then it is in the other's player's interest to defect. The game's single Nash equilibrium is mutual defection. The prisoner's dilemma is thus an example of how Nash equilibria are not necessarily optimal strategies.

④ **How the prisoner's dilemma applies to the tragedy of the commons**
The tragedy of the commons is a situation in which individuals acting in their own self-interest goes against the common good of the group, depleting a shared resource. An office might provide coffee for its employees, collecting contributions through an honesty box. Everybody has the temptation of having their coffee without paying for it, but if nobody pays then the coffee will stop. The same situation might happen with people travelling on public transport without buying a ticket, people illegally downloading films or music, a company over-fishing in a river, or a country producing a lot of pollution. Organizations generally try to overcome the problem by imposing punishments on defection to make it a less attractive strategy. Such punishments can include jail, fines, limiting of resources or social ostracism.

⑤ **How the prisoner's dilemma affects business, sport and diplomacy**
Picture two greengrocers opposite each other. If one of them starts selling avocados cheaper than the other one, they will win all the business and make a profit. But if the second business lowers its price to compete, then neither business has the advantage and both are making less profit than before. A similar situation happens with doping in sport: a single player has an incentive to take drugs to help them win a competition, but if all the players are taking drugs then nobody is better off than before and they have the added worry that they will be disqualified for cheating. A more powerful example is an arms race between countries. If one country has nuclear weapons and another does not, then they have an advantage, but if everybody has them then it is a costly situation in which nobody wins.

6 How to play the iterated prisoner's dilemma

It is interesting to see how players behave when the game is played multiple times and a running total is kept, since the advantages of mutual cooperation are enhanced and it is possible to build up trust between players. That said, if the game is played a fixed number of times, the strategy is no different to that of the one-shot game. Each player will reason that on the last round they should defect, using the logic from before. But then, since the outcome of the last round is predictable, they will then reason that they should also defect on the second-to-last round and so on, giving a strategy of defection on every round. So iterated versions of the game are either played over a very large number of rounds, or with the number of rounds unknown to both players.

7 The outcome of Axelrod's 1980 competition

In 1980, an American political scientist called Robert Axelrod organized a prisoner's dilemma tournament. Participants sent in computer programs and every program played off against every other program to see which performed best over about 200 iterations. The results were very different to those predicted by the single prisoner's dilemma. The highest-performing programs were those that were 'nice'; that is, they never defected first. However, some level of retaliation was important because a strategy that always cooperated would be taken advantage of by programs that always defected. The winner of the tournament was a simple four-line code submitted by Anatol Rapoport, called 'tit for tat', which started with cooperation and subsequently played whatever strategy their opponent had used on the previous round.

8 Lessons to be learned from the prisoner's dilemma

Axelrod had four pieces of advice based on successful programs in his prisoner's dilemma tournament. These were: be nice (don't defect first), retaliate (punish defection), forgive (return to cooperation if the other player does), and don't be envious (don't try to win more than your opponent). A strategy like tit for tat will never score more points than an opponent, but will come out ahead when playing many games over a variety of opponents. Similar strategies, such as 'tit for two tats', where the program only defects after two consecutive defections, also did well in the competition, since they did not get locked into alternating rounds of defection and cooperation.

9 What the best strategies are under Darwinian selection

The success of a strategy depends upon its environment. Is it possible that nice strategies do well against other nice strategies, but are exploited in a world of greedy ones? Axelrod performed an 'ecological' tournament in which successful strategies 'reproduced' and went forward to the next round, so that the number of each type of strategy depended on their success in the previous round. He found that a handful of nice strategies came to dominate the tournament, even if the tournament started with a predominance of greedy strategies. The altruistic (but retaliatory) strategies did well enough with each other to compensate for occasional losses, and would always grow in number, leading to the exploitative strategies dying out as they were out-performed.

10 What the prisoner's dilemma teaches us about morality

Certain aspects of human behaviour are seen across the world in different cultures throughout history. These include concepts of morality and justice, and are regulated in different ways, such as through the legal system, religion or social networks. Game theory argues that valued traits, such as fairness, kindness, humility and forgiveness, have emerged as the best strategies for the success of a group of individuals, showing that altruism is, in some sense, a form of selfishness. Altruism is seen in other animal species too, especially within kin groups, such as vampire bats sharing their food when others are hungry, monkeys grooming each other, meerkats standing guard against predators, and dolphins helping other injured dolphins.

TALK LIKE A GENIUS

❝ The British television show *Golden Balls*, hosted by Jasper Carrott, always ended with a prisoner's dilemma. Contestants who had worked together to accumulate a jackpot were reduced to two players, who were then given two golden balls labelled Split or Steal. If they both chose Split, they each got half the prize money, if they both chose Steal, they would leave with nothing, but if one player chose Steal while the other Split, then the stealer would win all the money. Over the three years the show ran, the contestants chose to split 53% of the time. Young men cooperated less than young women, with the effect reversing as the contestants got older. ❞

❝ In 2004, the University of Southampton was the first to win a prisoner's dilemma tournament without using 'tit for tat'. They submitted 60 programs labelled as either 'masters' or 'slaves', whose first objective was to determine if they were playing against one another or an outsider. A slave would cooperate with a master but defect against any outsider, while a master would always defect. This gave the University the top three results in the competition – along with many of the worst ones. ❞

I TRUE – If the second shop maintains its price, the first gains by undercutting. But if the second shop reduces its price, it is better for the first to do the same, to stay competitive. So reducing prices is the better strategy in each case.

2 FALSE – In the prisoner's dilemma, defection gives the best payoff, regardless of what your opponent does; yet mutual cooperation results in higher winnings over time.

3 TRUE – The advantages of mutual cooperation are enhanced when playing a game multiple times.

4 FALSE – Strategies that are always 'nice' and do not punish defection are taken advantage of by greedy strategies in games like the prisoner's dilemma.

5 TRUE – Winning strategies in large tournaments always had those traits, while greedy and exploitative strategies tended not to succeed.

THE BLUFFER'S SUMMARY

The prisoner's dilemma is a paradox in which each individual is motivated to act selfishly, but the best outcome comes from a group working together.

Probability

'An ingenious web of
probabilities is the surest screen
a wise man can place between
himself and the truth.'

GEORGE ELIOT, *ADAM BEDE*

Probability, the study of chance, was invented by
men wishing to become better gamblers. While it
is still used by casinos to ensure profits, probability
has had a far greater impact on the world than just
lining the pockets of poker players. Governments,
businesses, scientists, bankers and doctors all rely
on the mathematics of chance and risk, whether
it is deciding the value of a warranty or insurance
premium, choosing to vaccinate a population
against a possible epidemic, or agreeing whether we
have discovered a new kind of particle in physics.

**Being able to quantify our uncertainty about
the world has revolutionized science, allowing
us not only to understand the present, but to
predict the future.**

1 The probability of an event is the
count of the number of possible
ways it could happen.

TRUE / FALSE

2 To find the chance of two events
both happening, we multiply the
individual probabilities together.

TRUE / FALSE

3 It doesn't make sense to ask for
the chances of there being exactly
1.2mm of rainfall tomorrow.

TRUE / FALSE

4 When talking about the 'average'
value of a set of data, this always
means the data point that is exactly in
the middle of all the others (so half are
bigger and half are smaller).

TRUE / FALSE

5 When rolling two dice and adding
up the answers, a sum total of 7 is
more likely to arise than a total of 4.

TRUE / FALSE

TEN THINGS A GENIUS KNOWS

① How to measure probability
The probability of something happening is calculated by dividing the number of ways the event could happen by the total number of possible outcomes. So the chance of pulling an ace out of a properly shuffled deck of cards is 4/52, because there are four aces in the deck and there are 52 cards in total. The probability of an event A happening is written as P(A), and is always a number between 0 and 1. If the probability is 0 then the event is impossible, while if the probability is 1 then the event is certain to happen. The sum of all the probabilities in an event space is always equal to 1. This is like saying that if we draw a card from a deck, it must be equal to something.

② How to combine probabilities
When there are two events that we are interested in, A and B, we may wish to know the chance that both of them happen at the same time (the probability of A and B) or we may wish to know the chance that at least one of them happens (the probability of A or B). The first of these is written P(A∩B) and is calculated using multiplication, while the second is written P(A∪B) and is calculated using addition. So the chance of drawing an ace and then flipping heads on a coin would be 4/52 × 1/2 = 2/52, while the chance of drawing an ace or a king from a deck of cards would be 4/52 + 4/52 = 8/52.

③ What mutually exclusive events are
When adding probabilities, care needs to be taken to adjust the calculation if the events are not mutually exclusive. Mutually exclusive events are those that cannot happen simultaneously. A card drawn from a deck cannot be both an ace and a king, so in the example above it was correct to add the probabilities. But to find the chance of choosing either an ace or a heart from the deck, we need to take into account that the card could be the Ace of Hearts and subtract the possibility of this from the calculation. The general formula is P(A∪B) = P(A) + P(B) − P(A∩B).

④ The importance of independence
In 2010, a newspaper ran an article in which a woman was preparing scrambled eggs for breakfast and found that every single one of her six eggs was double-yoked. Since the chance of a single egg being double-yoked was 1 in 1000, the newspaper claimed that such an event was a 1 in a trillion phenomenon. But this would only be the case if each of the eggs being double-yoked was an independent event. That is, the chance of one egg being double-yoked was not influenced by the others being double-yoked. This was not the case: double-yoked eggs are larger than usual and are often laid by younger hens, meaning that a machine sorting eggs by size and farm are likely to group double-yoked eggs together. When events are not independent, the probabilities cannot simply be multiplied together and, instead, require a more complex calculation.

⑤ The difference between discrete and continuous probability
If it is possible to count all of the outcomes in a situation, as with a deck of cards or the tosses of a coin, then this is analysed using discrete probability theory. But there are many situations where this is not the case, such as predicting the amount of rainfall or the height of somebody chosen at random from a population. Here, it does not make sense to ask for the probability that a person is exactly 165.27 cm tall, so instead we need to use ideas from continuous probability theory. A more appropriate question would be 'What are the chances that a person is between 165 cm and 169 cm tall?', and to answer this we need techniques from calculus.

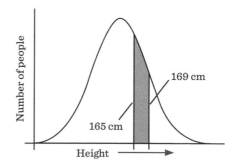

6 **What a probability density function is**
In continuous probability, the probability density function is used to describe the probability of a variable falling within a particular range. It is not the height of the graph at each point which gives the probability, but the area beneath it. This is found using the technique of integration, and is the continuous equivalent of summing the probabilities of different events. The total area beneath the graph must be equal to 1, just as the sum of probabilities in a discrete distribution must be equal to 1.

7 **How to measure averages and the spread of data**
When analysing a probability distribution, the two most frequent questions are: what is the average outcome, and how spread out are the values around this average? There are different kinds of averages: the mode describes the most common outcome; the median describes the middle value (so that 50 per cent of the outcomes fall below this value, and 50 per cent are above it) and the mean is a weighted average (the sum of all the values divided by how many there are). The variance measures the spread of the values and is the sum of the squares of the differences of each data point from the mean. The standard deviation is then the square root of the variance.

8 **What the normal distribution is**
Probably the best-known and most-studied probability distribution is the normal distribution, also known as the Gaussian distribution or bell curve. It has a very distinctive shape, in which the values are spread out symmetrically around a central point. This central point is the mean of the data. The example of the heights of the UK adult female

population is a normal distribution with a mean of 161 cm and a standard deviation (denoted by sigma, σ) of 6 cm. In any normal distribution, 68% of the data will lie within one standard deviation from the mean, 95% will be within two standard deviations, and 99% within three standard deviations.

9 **Where the normal distribution is seen**
The normal distribution arises in many different areas, though often after a scaling of the data. In biology, the logarithms of the lengths of living tissues tend to follow a normal distribution, and in finance the logarithm of exchange rates, stock prices and profits are assumed to be normally distributed. Exam scores are usually adjusted so that they follow a normal distribution, which is why students can find that their final mark is different from their raw score. In scientific experiments, the normal distribution is especially important because it is used to model the errors in measurements, under the assumption that a measurement is just as likely to be a little low as a little high from the true value.

10 **How the normal distribution affects all probability**
The rolls of a die are not normally distributed: every outcome has an equal chance of occurring. Yet something interesting happens when a die is rolled many times. A result called the central limit theorem says that the sum of all the dice rolls will approximate a normal distribution, with the approximation being more accurate the more rolls we make. And there is nothing special about dice: such a result is true for any collection of independent events with the same distribution, and, in particular, if an action or experiment is repeated many times. The reason for the central limit theorem is beautifully demonstrated by a Galton board (right) where balls fall through a grid of pegs. More balls will end up in the centre because there are more paths that lead to the centre than lead to one of the edges.

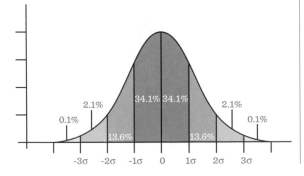

❝ To win money from your friends, invest in a set of non-transitive dice. These are three 6-sided dice, A, B and C, with the property that, on average, A beats B, B beats C and C beats A. So whichever die your friend chooses, you can choose one that will beat them! Even more amazingly, there is a set of five non-transitive dice so that if two friends each choose a die, you can always pick a third one that beats both of the other two. ❞

❝ When a major scientific discovery is made, such as the finding of the Higgs boson, you may read about sigma levels. Here sigma refers to the standard deviation, and is a measure of the confidence that the result was not a fluke. The Higgs boson is a five-sigma discovery, meaning that there is a one in three million chance that the observation happened by dumb luck. In medical studies, drugs must pass tests up to a two-sigma level of certainty. ❞

❝ The Gaussian distribution is found in many places outside probability theory, including in image processing where it is used to blur and smooth pictures. In a Gaussian blur, the pixel values of the original image are changed by a weighted average centred about a particular pixel, with the weights decreasing according to the Gaussian distribution as you move away from the pixel. The result is a smoothed image with reduced noise. ❞

1 FALSE – It is the number of ways the event could happen divided by the total number of possible outcomes, giving a number between 0 and 1.

2 FALSE – This is only true if the events are independent, so that the chance of one event happening does not influence the chance of the other happening.

3 TRUE – For continuous data, it makes more sense to ask for the chances of data falling within a range of values, e.g. the rainfall being between 1mm and 1.3mm tomorrow.

4 FALSE – There are different kinds of average, including the median (middle value), mode (most common value) and mean (weighted average).

5 TRUE – There are only three ways to get a total of 4 (1 + 3, 2 + 2 or 3 + 1), while there are six ways to get a total of 7 (1 + 6, 2 + 5, 3 + 4, 4 + 3, 5 + 2, 6 + 1).

THE BLUFFER'S SUMMARY

Probability is a measure of chance and takes values between 0 and 1, with 0 meaning that an event is impossible and 1 meaning that an event is certain.

Probability in the courtroom

'Under Bayes' theorem, no theory is perfect. Rather, it is a work in progress, always subject to further refinement and testing.'

NATE SILVER

Humans are particularly bad at making judgements concerning probability, and nowhere is this more important than in a courtroom, where the analysis may determine somebody's freedom or even their life. Bayes' theorem is a tool that can update probabilities based on new evidence, but its usage in court has been controversial.

Read the evidence yourself and see if you would find Bayes' theorem useful in determining somebody's innocence.

ARE YOU A GENIUS

1 Using a coin, the probability of flipping two heads and a tail (in any order), given that I have already flipped a head, is 1/2.

TRUE / FALSE

2 If 67% of criminals are illiterate, then 67% of illiterate people are criminals.

TRUE / FALSE

3 If you test positive for a rare disease, this means it is likely you have the disease.

TRUE / FALSE

4 If a DNA match is made that has a 1% chance of being found if the suspect is innocent, then their probability of innocence is 1%.

TRUE / FALSE

5 We cannot use probabilities to talk about past events, since either they have happened or they have not.

TRUE / FALSE

TEN THINGS A GENIUS KNOWS

1 What conditional probability is
In a courtroom, a defendant is assumed innocent until they are proven guilty. Each new piece of evidence that is presented will allow jury members to update their opinions on the likelihood of the person's innocence. What is the chance that they are innocent, given that they have an alibi for the night in question? What is the chance that they are innocent, given that DNA matching theirs was found at the crime scene? When we are considering the chance of something happening, given that another event has already occurred, this is called conditional probability. For a more numerical example, suppose that you win a bet if I flip two heads with a coin. To start with, your probability of winning is 1 in 4 (because there are four outcomes of flipping two coins: HH, HT, TH, TT). If I tell you that my first flip was a heads, your probability of winning changes to 1 in 2, but if you knew that I flipped a tails on the first toss then the bet is definitely lost.

2 What confusion of the inverse means
It is a common fallacy to confuse the probability of event A given event B, with the probability of event B given event A. Suppose you are gazing out onto a field and you see an animal with four legs. The chance that this is a sheep may be 25% (as it could also be a cow, pig or horse). But this does not mean that the chance of a sheep having four legs is 25%. Another classic example is from the Frenchman A. Taillandier, who, in 1828, found that 67% of criminals were illiterate, and concluded that illiteracy caused criminality. The chance that a criminal is illiterate is not the same as the chance that an illiterate person is a criminal. Mixing up these two probabilities is called confusion of the inverse.

3 The statement of Bayes' theorem
Bayes' theorem, named after Reverend Thomas Bayes, is a formula that allows people to convert between inverse conditional probabilities. Mathematicians use the shorthand P(A) to mean 'the probability of event A happening', and P(A|B) to mean 'the probability of event A happening, given that event B has already happened'. Using this terminology, Bayes' theorem is the following: P(B|A) = P(B) × P(A|B) / P(A). We can see this in action in the case of the coin-tossing game. We want to work out the probability of flipping two heads, given that the first toss is heads. So A = 'first toss is heads' and B = 'both tosses are heads'. P(B) is 0.25 (that is, 1 in 4). P(A|B) is 1 (that is, it is certain that the first toss is heads if we know that both tosses were heads). And P(A) is 0.5 (that is, 1 in 2). Putting this together gives P(B|A) = (0.25 × 1) / 0.5 = 0.5 or 1/2, as we expected.

4 The chance that an illiterate person was a criminal in 1828
How would we use Bayes' theorem to refute Taillandier's conclusion that illiterate people are criminals? To work out the true statistic, we need to work out the probability that someone is a criminal, given that they are illiterate. Bayes' theorem tells that P(criminal | illiterate) = P(criminal) × P(illiterate | criminal) / P(illiterate). That is, we need the additional information of the proportion of the population who are criminals, and the proportion of the population who are illiterate. Let us guess that, in 1828, this first number was about 5%, and the second was about 30%. Putting the numbers into the formula gives us the chance of being a criminal if you are illiterate as 0.05 × 0.67 / 0.3 = 11%. So although 67% of criminals were illiterate, only 11% of illiterate people were criminals.

5 How to interpret positive test results
Another use of Bayes' theorem is in working out the chances that someone has a disease, given that a test result has come back positive. To get the answer, Bayes' theorem says that we need to know the chances that the test would be positive if they really did have the disease (this is called the *sensitivity* of the test, and is hopefully very high); the chance that they would test positive whether or not they had the disease; and the proportion of the population with the disease. If it happens that the disease is very rare, then a false positive may be more likely than a true positive.

6 What the prosecutor's fallacy is
A version of the confusion of the inverse takes place in a courtroom, where it is often called the prosecutor's fallacy. Suppose we have found a piece of evidence that would be very unlikely if the defendant were innocent, such as a DNA match, or matching

shoe prints. The prosecutor's fallacy is to conclude from this that the chance of the defendant being innocent is also very unlikely. But the probability of the evidence being found if the defendant were innocent is not the same as the probability of the defendant being innocent, given that the evidence has been found.

7 Why lessons need to be learned from the Sally Clark case

A famous example of the prosecutor's fallacy is the case of the British mother Sally Clark, who was convicted in 1999 of murdering both her children. Statistics showed that the chance of one child (of the relevant age and background) dying of Sudden Infant Death Syndrome was 1 in 8500. Paediatrician Roy Meadow had assumed that the chances of two such deaths in the same family were independent, and so had multiplied this figure by itself to get a chance of 1 in 73 million for the two deaths. Not only was this assumption incorrect, but the conclusion – that Sally Clark must be guilty – failed to take into account that a double murder was even more unlikely.

8 Why the rarity of DNA may not be proof of guilt

Finding a match to a DNA sample taken from a crime scene may be rare: it may happen with only 1 in 10,000 people in the population. But if, say, 50,000 people in a city are all tested for a match, then we would expect about five matches to occur by chance alone. If the defendant were arrested before the DNA test was done, because of other suspicious behaviour, then the result may be significant. But if the defendant was identified to the police only after the DNA match in the population were found, then the culprit could just as easily be one of the other four people with a match. It would certainly be wrong in this example to claim that there is a 1 in 10,000 chance of the suspect's innocence, because there was a 1 in 10,000 chance of the DNA matching.

9 How Bayes' theorem is used directly in the courtroom

Bayes' theorem was used directly in a courtroom in the case of *R vs. Adams* in 1996. Denis Adams was on trial for the rape of a woman. Although he had an alibi for the night in question and did not match the victim's description of her attacker, there had been a match of his DNA – a match that the prosecution claimed had a 1 in 200 million chance of occurring. The jury were shown how to use Bayes' theorem in both the original trial and the later retrial, via a questionnaire that asked them questions like 'If he were the attacker, what is the chance that the victim would say he looked nothing like the attacker?'. Adams was convicted both times. In a second appeal, the Appeal Court was highly critical of the use of Bayes' theorem and the way that probabilities had been explained.

10 Whether Bayes' theorem has been banned

In 2010, a convicted killer known as 'T' appealed against the court's ruling, and won. The main evidence against him had been that a shoe print at the crime scene matched those of his Nike trainers. The prosecution had argued that this match was very unlikely. They had analysed the probabilities by looking at the number of pairs of trainers distributed in the country, the number of different sole patterns, the size of the shoe and so on, using Bayes' theorem to estimate the chances that the print at the crime scene came from the defendant. The Court of Appeal not only overturned the conviction, but also decreed that Bayes' theorem should not be used in court unless the underlying statistics were 'firm'. A judge in a different case claimed that it was impossible to assign probabilities to events that have already happened, but of which we are ignorant of the result, claiming that 'either something has happened or it has not'. This questions the very foundation of Bayesian probability, which allows us to update our view of events when we have new information about them.

❛ Bayes only published two scholarly works during his lifetime, and neither were about probability. One was about divine providence and the other was a defence of Newton's calculus. Bayes' theorem was only published two years after his death and was significantly edited by his friend Richard Price. ❜

❛ Bayesian probability is a different way of thinking about probability. Rather than thinking of the concept of probability as an estimate of the frequency of something happening, it becomes a quantity representing a state of belief. ❜

❛ Sally Clark was initially convicted for the murder of her two children. After two appeals she was exonerated, having served over three years in prison. Clark's experience has since been called 'one of the great miscarriages of justice in modern British legal history'. Roy Meadow, the former Professor of Paediatrics who mistakenly testified about the odds of double cot death, was struck off the medical register, although he appealed and was reinstated a year later. Three other women convicted on the basis of his evidence also had their convictions overturned. ❜

WERE YOU A GENIUS?

1 TRUE – If I have already flipped a head, then I still need to flip a head and a tail. The possibilities for two coins are HH, HT, TH, and TT, and 2 of these 4 are a head and a tail.

2 FALSE – This is called 'confusion of the inverse'. To calculate the true answer we would need to know what proportion of the population were criminals and what proportion of the population were illiterate.

3 FALSE – It depends on the accuracy of the test and on the rates of false positives. In a very rare disease, a false positive is more likely than a positive.

4 FALSE – If 100 random people are tested, a match is likely. The probability of innocence depends on more than just the DNA match probability.

5 FALSE – While this was the view of one judge, it is not the consensus among mathematicians, particularly those who are experts in Bayesian probability.

THE BLUFFER'S SUMMARY

Bayes' theorem helps prosecutors and defenders to find the chances of someone being innocent, given the evidence that is found against them.

Randomness

'We need myths to get by. We need story; otherwise the tremendous randomness of experience overwhelms us.'

ROBERT COOVER

Humans are not comfortable with randomness. We like to find rules and patterns in everything we encounter, even when they are not there. We see faces in random clouds, read the future in the random dregs of tea leaves and imagine we can predict the next random spin of the roulette wheel. But randomness is an inherent part of the universe and, paradoxically, understanding it better is the key to understanding our world and making predictions about the future. Randomness is how Einstein worked out the size of atoms. It is how the latest artificial intelligence is beating us at games. It gives us a way find the digits of π. And, of course, it explains why casinos will always make money.

Randomness, by its very nature, seems like something we can never understand, but geniuses over the years have harnessed it into a powerful mathematical tool.

1 If a coin has flipped heads a number of times in a row, then it increases the odds that the next toss will show a tails.

TRUE / FALSE

2 A gambler with a finite amount of money who keeps on playing a game with 50/50 odds will eventually lose all their money.

TRUE / FALSE

3 A drunk person wandering at random through a city will eventually return to the place they started.

TRUE / FALSE

4 Dust particles appear to move at random because they are buffeted by billions of collisions with air particles.

TRUE / FALSE

5 If casinos had no limits on betting, it would be possible to ensure a guaranteed profit in a game of chance.

TRUE / FALSE

TEN THINGS A GENIUS KNOWS

❶ How to lose money in a casino
In 1913, gamblers lost millions at the Monte Carlo casino in Monaco. The roulette ball was falling on black an unusual number of times in a row and, as it did so, more and more people put their money on the table, convinced that red was 'overdue' to come up. After the black had fallen 26 times many people were bankrupt. The gambler's fallacy, now also known as the Monte Carlo fallacy, is that the future of random events can be predicted from the past and that if something has happened less frequently in the past, then it will happen more frequently in the future. But the roulette ball has no memory of the past.

❷ Why a persistent gambler will always go broke
Imagine a coin-tossing game where flipping heads wins you £1 and flipping tails loses you £1. On average you would expect to win as often as you lose, but is it possible that if you had a lucky run of heads in the beginning you might come out ahead? Sadly not. Mathematics says that any gambler with a finite amount of money, however large, will always go broke playing this game. The result is called the 'gambler's ruin' and applies to any game with fair odds, where the gambler is playing against an opponent with infinite wealth (which essentially describes any casino).

❸ What a random walk is
The coin-tossing game above can be mathematically modelled as a 'random walk' on the number line. Starting at zero, you move one integer to the right if you flip heads and move one integer to the left if you flip tails. The gambler's ruin then predicts that the walk will always bring you back to the beginning in a finite amount of time. More than that, it predicts that a random walk will visit every integer an infinite number of times, so with unlimited resources you would be guaranteed to win £1 million at some point. A general random walk is any path in which the direction of movement is decided according to a fixed set of probabilities.

❹ Why a drunken wanderer will always return home
Random walks can happen on other, more interesting, spaces than the number line. A drunk mathematician emerging from a pub may go on a random walk in the city, making a random choice at each intersection about which way to go. This is a 2-D random walk, and theory predicts that the drunkard will always find their way back to their original pub. A drunken bumblebee will not be so lucky. Random walks in three dimensions are only 34% likely to return to the start. Simulations of random walks have been used to model fluctuations in stock markets, grazing paths of animals, and the way people move between websites online.

❺ What Brownian motion is
In 1827, Robert Brown looked down a microscope at pollen grains in water. The pollen grains appeared to be moving of their own accord, dancing around as if they were alive. After finding the same behaviour in dust particles,  Brown concluded that he had not found a new form of life, but a new scientific phenomenon. A number of people tried to explain it, but it was Einstein who found the answer. The dust and pollen particles were undergoing a random walk, buffeted by the motions of the billions of surrounding water or air molecules. Today, this is called Brownian motion. In studying how it worked, Einstein was able to calculate the size of atoms and work out a value for Avogadro's number: how many atoms are in a mole of a gas.

❻ What a stochastic process is
The movement of a pollen grain in water cannot be predicted exactly. It undergoes something like 100 million million collisions every second, as do all the particles in the air around us. Instead, scientists model the particles using a stochastic process, which is a system governed by probability. Stochastic modelling harnesses the underlying randomness in a system and uses it to make predictions about large-scale outcomes. For example, models of air molecules

colliding randomly with each other create accurate predictions about the pressure and temperature in a room. Modelling the randomness in stock markets can make predictions about optimal pricing. And modelling the randomness of radioactive decay informs the design of nuclear power stations and of radiotherapy treatment for cancers.

7 **Why card-counting is banned in casinos**
Brownian motion and random walks are both examples of Markov processes. These are stochastic processes with the property that the future is only determined by the present and not by the past. The coin you are flipping does not remember all the ways it has already fallen. A roulette wheel is a Markov process, which is why casinos will happily publish data on all the past spins of a wheel, hoping that the gambler's fallacy will cause players to bet irresponsibly. But they will not do so on a blackjack table, because the cards that are yet to come up are influenced by what has already been on the table. This is why card counting is banned, because prior knowledge of the game gives gamblers an advantage over the dealer.

8 **Why casinos put limits on bets**
In 18th-century France, there was a special betting strategy called a martingale. In a coin-tossing game, the gambler would first bet £1. If they lost, they would bet £2 on the next toss, then £4, then £8, doubling their bet each time. Whenever they finally won, the winning amount would counteract all previous losses, ensuring the gambler always won their original £1 stake. Casinos soon caught onto this, putting limits on the bets that could be placed. Today, the word martingale is used to describe a type of stochastic process where the expected winnings on the next event is the same as the expected winnings on the current event. Martingales model fair games where predictions of future winnings do not depend on the winnings or losses of the past. This sounds very similar to a Markov process, but the two are different. It is possible for a situation to be a martingale without being Markov, because knowledge of past events may influence the probability of winning, even though it does not influence the value of the win.

9 **How to find the value of π using randomness**
Random processes can be used in clever ways to find answers to questions that are not random at all. Buffon's needle is an example of this. Draw a set of parallel lines on a sheet of paper, a needle's width apart. Then drop needles onto the paper and count how many needles cross over a line. The total number of needles dropped, divided by the number of needles that cross a line, is an estimate for half the value of π, and this estimate gets better and better with the more needles that are dropped.

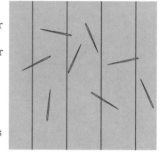

This surprising result comes from considering the number of ways that a needle might cross a line, which is a combination of the angle of the needle and the closeness of the centre of the needle to a line. Calculus (integration) of the sine function is required to get the answer, which is what produces the value of $\pi/2$.

10 **What Monte Carlo methods are**
Buffon's needle is an example of a Monte Carlo method, which is a way of doing a large number of experiments and using the data to determine exact results. The technique is especially helpful when it is too difficult to solve an equation exactly, which happens often when we model the complicated realities of the world. It was developed during the cold war by Stanislaw Ulam and John von Neumann, who used it to work out the numbers necessary for developing the hydrogen bomb. Today, it is used in everything from modelling the evolution of galaxies to determining the reliability of new bridges. It tells search-and-rescue parties where to look after a plane crash, without having to solve the equations for ocean currents. And it is enabling artificial intelligence to beat the best Go masters, by simulating the outcomes of many different games in order to find the best moves.

TALK LIKE A GENIUS

❝ Generating true randomness is surprisingly difficult, for both humans and computers. Nowhere is this more important than in cryptography. The Enigma machine in the Second World War would have been an unbreakable cipher, had it not been for the operators setting up the machines each day. Instead of using a random 3-letter code, they would often use their name or the initials of their sweetheart, allowing mathematicians at Bletchley Park to crack the code. Even today, the problem is not solved; pseudorandom number generators are used to pick primes for encryption, but often they are not random enough, and allow a clever eavesdropper to decrypt the messages. ❞

❝ It is not only humans who like to see patterns where in reality there is only randomness. In 1947, B F Skinner demonstrated that pigeons were superstitious animals too. Skinner put hungry pigeons in a cage where food was released by a timer operating entirely at random. The pigeons soon developed strange behaviours, such as turning around three times or tossing their head from side to side. This was because they were imitating the behaviour that they happened to be doing at the moment the food was released, in the hope that it was the cause of the mechanism. ❞

WERE YOU A GENIUS?

1 FALSE – Such a belief is called the gambler's fallacy. The next toss of a coin is completely independent from the tosses that have been made previously.

2 TRUE – This is called the gambler's ruin.

3 TRUE – A random walk in two dimensions is almost certainly (that is, with probability 1) guaranteed to return to the starting point within a finite time.

4 TRUE – The random walk of the dust particle is called Brownian motion.

5 TRUE – In a game with 50/50 odds, you can ensure a guaranteed win of your initial stake by doubling your bet each time you play.

THE BLUFFER'S SUMMARY

Stochastic analysis uses our understanding of uncertainty to make accurate predictions about the future.

Paradoxes of probability

'The 50-50-90 rule: anytime you have a 50-50 chance of getting something right, there's a 90% probability you'll get it wrong.'

ANDY ROONEY

We all think we understand the ideas of chance, but most of us are fooled by simple paradoxes of probability. Our bad intuition for chance explains why we play the lottery, even though we are less likely to win than to die before the draw happens. It explains why we marvel at coincidences, even when they are not so improbable. It is the reason why casinos will always make money and people will lose on gameshows. But strange probability facts are also helping us to spot corruption in banking and voting.

Most people's instincts about chance are too strong to overcome any mathematical training. Are you genius enough to avoid the trap of a probability paradox?

1 The odds of choosing six numbers correctly out of 59 is roughly 1 in 10.

TRUE / FALSE

2 With only 23 people in a room, it is likely that two people will share the same birthday.

TRUE / FALSE

3 Someone secretly flips two coins and tells you that at least one coin shows 'heads'. The chances that the other coin also shows heads is 50%.

TRUE / FALSE

4 If a deck is perfectly riffle-shuffled eight times it returns to the same order it started in.

TRUE / FALSE

5 The leading digit of a random number chosen from a newspaper is six times more likely to be a 1 than a 9.

TRUE / FALSE

TEN THINGS A GENIUS KNOWS

1 **Why we never win the lottery**
To win the UK national lottery, a player has to correctly pick 6 numbers out of 59. This does not seem so implausible at first glance. After all, 6 out of 59 seems roughly like 1 in 10. Sadly, these are not the odds of winning. How many ways can we choose 6 numbers out of 59? There are 59 choices for the first number, 58 for the second and so on, down to 54 choices for the sixth number. This gives us 59 × 58 × 57 × 56 × 55 × 54, which is about 32 billion combinations. Since we don't care what order the balls are chosen, we divide by the number of ways of arranging six numbers (which is 720), to find that there are about 45 million ways of choosing the lottery numbers, only one of which is correct each week. But if 45 million people play the lottery each week, there is a high chance that *somebody* will win.

2 **How many people need to be in a room before two share a birthday**
The birthday paradox asks how many people need to be in a room before it is likely that two people will share a birthday. Most of us reason that there are 365 possible days in a year, so we would need 365/2 = 183 people before the odds are in our favour. But, similarly to the lottery calculation, the correct thing to consider is how many possible pairings of people there are. This is because we are not concerned with matching a particular birthday, but finding any pair of people with a matching birthday. With only 10 people in a room, there are already 55 possible ways to pair them up. It turns out that with only 23 people in a room, there are so many possible pairings that the probability of two people sharing a birthday goes over 50%.

3 **Why we are surprised by coincidences**
We tend to overestimate the likelihood of winning the lottery because we confuse the chances of *us* winning with the chances of *somebody* winning. This phenomenon works in reverse with coincidences, where we underestimate the chances of them happening because we are focused on the specifics. For example, we might be thinking of a person and then at that moment they happen to ring us. What were the chances of the person calling at exactly the same time as we were thinking of them?

Were they psychic? Probably not. When we think of all the possible friends that we have and all the phone calls we receive, and the fact that we never stop to count the number of times we have thought of a friend and they haven't called, the chances are high that this particular coincidence will happen at some point, even though the chances of it happening at that specific moment are small.

4 **What the Monty Hall problem is**
In the American game show *Let's Make a Deal*, host Monty Hall presented a player with three doors. Behind one of the doors was a car but behind the two other doors were goats. The player initially chose a door, then Monty opened one of the two other doors to reveal a goat. Finally, the player was given a chance to stick with their original choice or to switch to the other unopened door. Is it better for them to stick or switch, or is there no difference? Most people will argue that since there are two unopened doors, there is a 50-50 chance of the car being behind each one, and therefore it makes no difference whether the player sticks or switches. Paradoxically, the player can double their odds of getting the car by switching to the other door. This is because the chance of them picking the right door in the first place was 1 in 3, so the chance of the other door being the right one is 2 in 3.

5 **The boy or girl paradox**
One day I am at a party and I meet a woman who tells me that she has two children. She later reveals that she has a son. What are the chances that her other child is also a boy? Our instinct is to reason that the gender of her other child is independent of the fact that she has a son, and is thus equally likely to be a boy or a girl. In fact, her other child is twice as likely to be a girl than a boy. We get this result by considering all the possible combinations of genders of two children. Writing B for a boy and G for a girl, we could have BB, BG, GB or GG. We know she has at least one son, which excludes the GG option. Of the three remaining options, two of them have the second child as a girl while only one has the second child as a boy.

6 **The Tuesday boy problem**
If the boy or girl paradox was not strange enough, there is an extension to the problem that seems incomprehensible. In the Tuesday boy problem, the woman at the party reveals that she has a son born on

a Tuesday. What are the odds that her second child is also a boy? Once again, our instinct reasons that the day of birth cannot possibly be useful information in calculating the odds of the gender of the second child. However, writing out all the options of gender and day of birth where at least one child is a boy born on a Tuesday, we discover that the odds of the second child being a boy are 13/27. Paradoxically, the more information is known about the first child, the closer the chances of the second child being a boy will get to 1/2, which was our instinctive answer.

7 How to win on the heads-flipping game
I ask you to choose a random combination of three heads or tails. For example, you might pick heads-heads-tails (HHT). I then choose my own random combination of three heads or tails. We start flipping a coin, and whoever's combination comes up first will win the round. Although the coin tosses are completely random, and every sequence of three outcomes is equally likely, it is always possible for me to choose a combination of heads or tails that is likely to come up before yours. My method is to use the first two options of your sequence as the last two of mine; then as my first flip I choose the opposite of your middle flip. So if you pick HHT, I pick THH. In this case my choice makes me three times more likely to win than you, and with some combinations I can increase my odds to 7-to-1. The idea is that I pick a combination that is going to beat you if you do not win immediately. If you flip T first, or HT, then you lose and must start again trying to flip two heads, but if you now flip two heads then THH has appeared and I win.

8 How to randomize (or not) a deck of cards
A standard deck of cards has 52 cards in it. The number of different shuffles of a deck is 52! = 52 × 51 × 50 × 49 ×. . ., which is a number with 68 digits – comparable to the number of molecules in the universe. A truly randomly shuffled deck is therefore unlikely to have ever been seen before in the history of humanity. But how easy is it to create a random shuffle? The consensus among mathematicians is that six or seven riffle shuffles, where two halves of the deck are roughly interleaved, is enough to randomize the deck. However, a skilled magician can use a surprising mathematical fact to cheat using this method. In a weave shuffle, two halves of the deck are perfectly interleaved – alternating between one half of the deck and then the other. It turns out that doing eight perfect weave shuffles, keeping the same half of the deck on top each time, puts the deck back in exactly the same order as in the beginning.

9 How Benford's law predicts more ones
When we look at the numbers between 0 and 99 we see that ten of them start with a 1, ten of them start with a 2, ten of them start with a 3 and so on, with every leading digit being equally likely. It seems like a paradox, then, that in real-life data there is a probability of over 30% that the leading digit of a number is 1 and 17% that it is a 2, but less than 5% of it being a 9. This strange finding is called Benford's law after the American physicist Frank Benford who investigated it in 1938, and it states that the distribution of leading digits follows a logarithmic curve rather than a linear curve. (This means the sequence of numbers decreases by successively dividing rather than subtracting.) The effect is most noticeable on data that spans several orders of magnitude, such as prices on a stock market or areas of lakes, rather than with narrow data with a clear maximum and minimum, such as shoe size or human height.

10 How to use Benford's law to fight corruption
Most people are unaware of Benford's law, so if they try to falsify data they make an effort to spread out the numbers equally among the leading digits. When real data is compared with made-up data, the difference becomes noticeable. There are many examples of fraud being detected using Benford's law, including embezzlement of company money, misleading tax returns, credit-card fraud, made-up election numbers, falsified economic data, price fixing, malicious cyber attacks and spam Twitter accounts.

TALK LIKE A GENIUS

❝ If you decide to play the lottery there are ways of choosing your numbers to avoid sharing your big prize with anyone else. Although all combinations of numbers are equally likely to win, some combinations are more popular among players. Be sure to choose some numbers above 31 to avoid the people who play using birthdays. Don't choose numbers all in the same row or column and don't be afraid to pick pairs of consecutive numbers, as people generally try to space their numbers out. ❞

❝ With only seven people in a room, the odds are that two people will have a birthday within a week of each other. ❞

❝ When Marilyn vos Savant, often called the smartest person in the world, posed the Monty Hall problem in a newspaper column in 1990, she received nearly 10,000 responses, with most people disagreeing with her answer. The responses included letters by professional mathematicians, saying things like 'There is enough mathematical illiteracy in this country, and we don't need the world's highest IQ propagating more. Shame!' ❞

❝ Benford's law was actually discovered by astronomer Simon Newcomb, who in 1881 noticed that in his tables of logarithms, the earlier pages that started with a 1 were much more worn than the other pages. ❞

WERE YOU A GENIUS?

1 FALSE – The odds are about 1 in 45 million, which is the number of possible ways of choosing 6 different numbers from 59.

2 TRUE – With 23 people the odds are above 50% that two will share the same birthday.

3 FALSE – The odds are only 1/3. The possible combinations are HH, HT and TH, and in only one of these are there two heads.

4 TRUE – A perfect weave shuffle, in which two halves of the deck are interleaved, returns the deck to its original order after eight passes.

5 TRUE – The leading digit has about 30% chance of being a 1, and less than a 5% chance of being a 9.

THE BLUFFER'S SUMMARY

Our intuition about probability is very often wrong, but overcoming our instincts and learning to apply mathematics can help us win bets, understand coincidences and even fight corruption.

Glossary

ALGEBRA
The field of mathematics involving symbols and the rules for manipulating them in equations and inequalities. 'Abstract algebra' embraces complex mathematical objects and links fields such as group theory and topology.

ALGORITHM
A recipe or set of instructions describing how to solve a particular type of mathematical problem.

ANALYSIS
The branch of mathematics concerned with the study of limits and, by extension, the behaviour of functions.

AXIOM
A statement that can be accepted without proof, because it can be taken as self-evident.

BASE
The number of different digits or other symbols used in a positional system of counting to represent numbers – for example, the common 'base 10' system uses digits from 0 to 9.

BINARY
The base-2 system of counting that uses just two digits, 0 and 1. Binary is widely used in modern digital electronics.

CALCULUS
A branch of mathematics that uses functions called differentiation and integration to study rates of change and gradients, and cumulative sums or areas, respectively.

CARDINALITY
A measure of the number of elements within a set or collection of items.

CHAOS
The branch of mathematics studying dynamical systems and equations that are highly sensitive to initial conditions, and can evolve to wildly different values from very similar original states.

COMPLEX NUMBER
A number with real and imaginary parts, typically expressed in the form $a+bi$, where i is the square root of -1 and a and b are real numbers.

CONSTANT
Any component of an equation that does not vary.

CONTINUITY
A property of 'smooth' mathematical functions – those in which small changes in input always produce small changes in output.

CONVERGENCE
The tendency of an infinite sequence to come closer to a value called the limit as the number of terms increases, or of a function to approach a limit as one of its variables increases.

COORDINATE
In geometry, a number used to describe the position of a point in a particular dimension.

COUNTABLE
Describing an infinity that is the same size as that of the natural numbers.

CUBE (NUMBER)
To cube a number is to multiply it by itself three times. Hence, 5 cubed is $5 \times 5 \times 5 = 125$.

CUBIC (POLYNOMIAL)

A polynomial of degree 3, e.g.
$x^3 + 3x^2 - 2x + 7$

CYCLIC GROUP

A group whose elements are generated by applying one symmetry repeatedly. For example, the four rotations of a square are generated by applying a rotation of 90° four times in succession.

DENOMINATOR

The number below or to the right of the line in a fraction, indicating the number of equal parts into which the numerator is divided. Sometimes also known as the divisor.

DERIVATIVE

A function that measures the sensitivity of another function to changes in its input variables – for instance, the gradient of a graph at a certain point.

DIAMETER

The furthest straight-line distance across a circle, passing through its centre and connecting two points on its edge.

DIFFERENTIATION

In calculus, the process used to find the derivative of a function – the rate of change in its output with respect to one of its inputs.

DIMENSION

A measure of the number of coordinates required to specify the location of a point in a mathematical space.

DIVERGENT

Descriptive of a sequence that does not converge on a limit. Divergent sequences may rise towards infinity at higher terms, or oscillate around a value while not approaching it as a limit.

e

Also known as Euler's number, e is a mathematical constant, the limit of the infinite series $(1+1/n)^n$ as n approaches infinity. With a numerical value of roughly 2.718, e appears in many widely separated fields of mathematics, especially calculus.

ELLIPSE

A closed curve on a plane, with two focal points and the property that the sum of the distances to these two points is the same for each point on the curve.

ELLIPTIC CURVE

A curve (not directly related to the ellipse) defined by an equation of the form $y^2 = x^3 + ax + b$, where a and b are numbers chosen so that $x^3 + ax + b$ has three distinct roots. Elliptic curves are surprisingly relevant to widely varied areas of mathematics.

EQUATION

A mathematical statement that expressions on either side of an equals (=) sign have the same value.

EXPONENT

The power to which a particular number or mathematical expression should be raised, usually indicated by a superscript to the right of the number.

FIELDS MEDAL

The equivalent of the Nobel Prize in mathematics, and the highest honour that a mathematician can achieve. It is only awarded to mathematicians under the age of 40.

FRACTAL

A mathematical object that is self-similar, revealing an infinite level of detail in recurring patterns at smaller and smaller scales.

FRACTION

A mathematical expression indicating the division of one quantity by another. The quantity to be divided (the numerator) is shown above or to the left of a line, with the number of divisions (the denominator) below or to the right.

FUNCTION

An expression that takes a number, or collection of numbers, as an input, and produces an output. For example,
$$f(x) = sin\ x$$
is a function, as is
$$g(n,m) = m + 2n.$$

GAME

In mathematical 'game theory', a situation in which two or more participants can compete or cooperate for resources, with the behaviour of the players resulting in varying rewards.

GEOMETRY

The field of mathematics concerned with the properties of space and the shapes, sizes and positions of points and objects within that space.

GOLDEN RATIO

If a line is cut in two so that the ratio of the long part to the short part is the same as the ratio of the whole line to the long part, then this ratio is called the golden ratio. It has a value of about 1.618.

GRADIENT

A measure of the steepness of a graph at a particular point, or more broadly the rate at which the output of a function varies with respect to changes in a particular input variable.

GRAPH

In the broad mathematical field of graph theory, a topological object consisting of nodes (points) and the edges connecting them. A graph can also mean a plot of points in two dimensions, often to represent the values of variables related by a function.

GROUP

In abstract algebra, a mathematical structure consisting of a set of elements along with an 'operation' that can be performed on them while obeying fundamental rules called the group axioms. The integers along with the operation of addition are a simple example, but groups can encompass a broad range of mathematical objects.

IMAGINARY NUMBER

Any number that is the square root of a negative number. Such numbers cannot be calculated directly using the mathematics of real numbers, but but can be thought of as lying on a separate number line perpendicular to the real numbers.

INEQUALITY

A mathematical statement of the relative values of two expressions, with the relationship indicated by symbols such as > ('greater than') and ≤ ('less than or equal to').

INFINITY

A mathematical quantity larger than any natural number and having no upper bound.

INTEGER

A positive or negative whole number.

INTEGRATION

In calculus, the process used to find the integral of a function – a property (typically the area it encloses on a graph) arising from its changing output across many infinitesimally small divisions.

IRRATIONAL NUMBER

A number that is not rational and cannot be expressed as a common fraction of two integers.

LIMIT

A value that a convergent sequence approaches at higher terms, or which the output of a convergent function approaches at higher values of an input variable.

LOGARITHM

The inverse of exponentiation. In base 10, the logarithm of 1000 is 3, because 10^3 is 1000. The natural logarithm uses the base of e and is written $\ln(x)$ or sometimes just $\log(x)$.

MATRIX

A grid of numbers, often representing a collection of vectors. There are special rules on how multiplication between matrices is performed.

NATURAL NUMBER

A positive whole number of the type that can be used for counting objects in the real world.

NUMERATOR

The number above or to the left of the line in a fraction, which is divided by the denominator. Sometimes also known as the dividend.

PI (π)

A constant, defined as the ratio of a circle's circumference to its diameter, but occurring in widely separated fields of mathematics. Pi is an irrational number with a numerical value of roughly 3.142.

PLATONIC SOLID

A regular polyhedron with a convex rather than star-like form. Only five such figures exist – the tetrahedron, cube, octahedron, 12-faced dodecahedron and 20-faced icosahedron.

POLYGON

In geometry, a figure on a flat plane bounded by a finite number of straight-line segments (at least three are needed in order to define the area of a figure). In regular polygons, all the line segments have the same length.

POLYHEDRON

In geometry, a three-dimensional geometric object with a finite number of flat polygonal faces. In regular polyhedra, all the faces and edges are identical in size and shape.

POLYNOMIAL

An expression in terms of the sum of powers of an unknown variable; for example, $5x^3 - 1/2\, x^2 - \sqrt{2}$. The numbers in front of the powers of x here are called the coefficients of the polynomial.

POWER (OF A NUMBER)

To raise a number to a certain power is to multiply it by itself that number of times. Superscripts are used to indicate the power, so 3^4 means the number 3, multiplied by itself four times, i.e. $3 \times 3 \times 3 \times 3$.

PRIME NUMBER

A natural number that is exactly divisible only by itself and 1. (1, however, is not a prime number.)

PROBABILITY

The branch of mathematics concerned with chance – the distribution of certain occurrences and the likelihood of particular outcomes in different circumstances.

PRODUCT

The result of multiplying one mathematical object with another (most commonly the multiplication of two numbers).

PROOF

A mathematical argument that shows the correctness of a statement through a sequence of logical deductions beginning with initial axioms.

PYTHAGORAS'S THEOREM

A fundamental relationship in Euclidean geometry, linking the three sides of a right-angled triangle (the hypotenuse, length a, and the other two sides, lengths b and c) through the equation $a^2 = b^2 + c^2$.

QUADRATIC

A polynomial of degree 2, e.g. $x^2 - 5x + 6$.

QUATERNION

A type of complex number extending to four dimensions, with one real component and four imaginary ones, expressed in the form $a + bi + cj + dk$.

RADIUS

Half the diameter of a circle – that is, the distance from its centre to any point on the edge.

RATIONAL NUMBER

A number that can be written as one integer divided by another.

REAL NUMBER

Any number that can be expressed in decimal form.

REFLECTION

In geometry, a transformation that maps points in a space to an identical distance across an axis or plane of reflection, creating a mirror-image of the original.

REMAINDER

In simple arithmetic, the amount 'left over' after dividing one integer by another and accepting only an integer result.

ROOT (OF A NUMBER)
The root of a number n is the number that produces n when it is raised to a given power (i.e. multiplied by itself a certain number of times).

ROOT (OF A POLYNOMIAL)
A number that makes a polynomial output zero. For example, $x = 2$ is a root of $x^2 - 5x + 6$ because $2^2 - 5 \times 2 + 6 = 0$.

ROTATION
In geometry, a transformation that turns an object or figure about a fixed point.

SEQUENCE
An ordered list of numbers, such as 1, 3, 5, 7, 9, ... Each number in a sequence is known as a term.

SERIES
The sum of an infinite sequence, achieved by adding up all of its terms. A series may have a finite value if its associated partial (finite) sums converge to a limit.

SET
A collection of unordered, distinct objects (such as numbers, functions, books or sheep). Sets may be finite or infinite. Members of a set are called elements.

SPACE
In classical geometry, a space is a region of two or more dimensions in which positions and directions of objects can be defined. In broader mathematics, it is defined as a set of mathematical objects to which some sort of structure is applied.

SQUARE (NUMBER)
To square a number means to multiply it by itself. A number is called square if it is the square of another number, e.g. 4, 9, 16.

SQUARE ROOT
The square root of a number n is the number that needs to be multiplied by itself to give an answer of n.

SYMMETRY
The property of invariance, whereby a mathematical object remains unaltered after a transformation of some sort.

TANGENT
A straight line that just touches another curve or surface at a point. The gradient of a tangent line is equal to the derivative of the curve at that point.

TOPOLOGY
The branch of mathematics concerned with the shape of objects and the properties that remain unaltered when they are subjected to continuous deformations.

TRANSFORMATION
Geometric transformations are operations such as reflection, rotation or translation that move or 'map' points or objects onto new locations. More broadly, a transformation may be any operation on a mathematical object or set of objects.

TRANSLATION
In geometry, a transformation that moves the position of an object without changing its size or orientation in other ways.

TRIGONOMETRY
The branch of mathematics specifically concerned with the properties of triangles.

UNCOUNTABLE
An infinity that is larger than that of the natural numbers.

VARIABLE
An unknown quantity, usually denoted by a letter like x or t.

VECTOR
A geometric object that has both a magnitude and a direction.

VERTEX
A corner of a polygonal shape, or the end point of a line.

Index

ABOUT THE AUTHOR

Julia Collins has a PhD in four-dimensional knot theory from the University of Edinburgh, where she spent five years as the Mathematics Engagement Officer, with a remit to lecture and spread an appreciation of mathematics. She is now the CHOOSE**MATHS** Women in Maths Network Coordinator at the Australian Mathematical Sciences Institute. Julia's writing has been published in Nature and in Princeton University Press' anthology *The Best Writing on Mathematics*. She is a winner of the How to Talk Maths in Public competition, has been nominated for the London Mathematical Society's Anne Bennett prize, and co-organized the world's first Maths Craft Festival.

ACKNOWLEDGEMENTS

Even with a mathematics degree and PhD, there are some chapters I could not have written without the help of a patient mathematician. The Hodge Conjecture was by far the most difficult chapter: thank you Geordie Williamson for taking the time to explain this to me. Thank you also Diarmuid Crowley and Andrew Ranicki for double-checking that I had written something vaguely sensible. Andrew Ranicki (who sadly passed away while this book was being written) deserves additional thanks for providing the MRI-scanning analogy, for providing interesting facts about Hodge and Weil, and for helping me to explain the Poincaré conjecture precisely. The next-hardest chapter was about the Yang-Mills and mass gap conjecture: thank you José Figueroa-O'Farrill for correcting my most egregious mistakes and for helping me to understand quantum field theories.

My greatest thanks must go to my two proof-readers, without whom this book would not be nearly as accurate and well-written. Thank you Ben Goddard, for always being there for me, no matter how busy you are, and for applying your wonderful pedantry to checking all the mathematics in the book. Thank you Douglas McQueen-Thomson for meticulously reading every chapter and for teaching me so much about being a better writer.

Thank you Brett Hale, for being there at the beginning and always believing in me. Thank you to my colleagues at the Australian Mathematical Sciences Institute, especially Inge Koch, for supporting me while I was writing this book.

Thank you to the editors at Quercus, for putting up with delayed deadlines and a never-ending list of edits, not to mention taking a chance on a new author. I very much appreciate the opportunity you have given me in writing this book.

Finally, thanks go to all the baristas around Australia and New Zealand who kept me caffeinated during the writing of this book, especially Carolina (Brunswick East, Melbourne) and the Ministry of Caffeine (Buderim, Queensland).

PICTURE CREDITS

First published in Great Britain in 2018 by

Quercus Editions Ltd
Carmelite House
50 Victoria Embankment
London EC4Y 0DZ

An Hachette UK company

A CIP catalogue record for this book is available
from the British Library

HB ISBN 9781786483355
EBOOK ISBN 9781786483362

Every effort has been made to contact copyright holders. However,
the publishers will be glad to rectify in future editions any
inadvertent omissions brought to their attention.

The picture credits constitute an extension to this copyright
notice.

10 9 8 7 6 5 4 3 2 1

Printed and bound in China